546-8

The Periodic Table of Elements Compendium

The Periodic Table of Elements Compendium

Compiled by Helen Eccles

Edited by Jacqui Clee and Neville Reed

Layout by David Clee

Cover designed by Imogen Bertin

Published by the Education Division, The Royal Society of Chemistry

Printed by The Royal Society of Chemistry

Copyright in this form The Royal Society of Chemistry 1994

This material may be reproduced without infringing copyright providing reproduction is for use in the purchasing institution only. The permission of the publisher must be obtained before reproducing this material for any other purpose. For further information on other educational activities undertaken by the Royal Society of Chemistry write to:

The Education Department

The Royal Society of Chemistry

Burlington House

Piccadilly

London W1V 0BN

ISBN 1-87-0343-27-1

British Library Cataloguing in Publication Data.

A catalogue for this book is available from the British Library.

Introduction

The information contained in this publication was originally supplied to NERIS (National Educational Resource Information Service) for access via the online system; and as facsimile records on CD-Rom. NERIS ceased to exist in April 1993 and the Royal Society of Chemistry decided to publish the records in a printed form to make them readily accessible to all educational institutions.

The Society is grateful to Robert Bailey and David Taylor at NERIS for their support and financial backing to initiate the original work; Dr Helen Eccles for producing the original records; and to David and Jacqui Clee for turning the original records into the final printed form.

Patterns in the Periodic Table

The Periodic Table is an arrangement of all the known elements in order of increasing atomic number. The reason why the elements are arranged as they are in the Periodic Table is to fit them all, with their widely diverse physical and chemical properties, into a logical pattern. If sodium is placed beneath lithium and not next door to fluorine, and potassium is placed beneath sodium to begin another row - and so on - it is found that the vertical lines of elements are chemically similar. These vertical lines are called GROUPS.

Horizontal lines of elements are called PERIODS. A set of D-BLOCK ELEMENTS, sometimes called the transition metals, occurs between Groups II and III; these are also chemically similar to each other. Some Groups exhibit striking similarity between their elements, such as Group I, and in other Groups the elements are less similar to each other, such as Group IV, but each Group has a common set of characteristics.

The Periodic Table is divided into BLOCKS.

The s-block elements have valence configuration s^1 or s^2.

The p-block elements have valence configuration s^2p^1 to s^2p^6.

The d-block elements have valence configurations in which d-subshells are being filled.

Hydrogen occupies a unique position at the top of the Periodic Table. It does not fit naturally into any Group.

All the members of a Group have the same valence configuration but different principal quantum numbers. The number of valence electrons equals the Group number. The Period number equals the principal quantum number of the valence shell.

Chemically, elements in the same block exhibit the same general characteristics. This is most apparent for the s-block elements which are all metals with low electronegativity. The p-block elements are more varied with some metals such as aluminium on the left and non-metals on the right. Between them, indicating the gradual change in character going across the Periodic Table are the metalloids (or semi-metals), which lie roughly in a diagonal line from silicon to tellurium. The d-block elements are often called the transition metals, but some of them such as zinc do not fit this description well. They are usually considered together as differences between Groups are much less apparent in this block.

Periodicity of some properties

Periodicity is the name given to regularly-occurring similarities in physical and chemical properties of the elements. Periodicity reflects the periodic repetition of similar electron configurations. Very many properties of the elements show periodicity. The most obvious of these is the continuing change from metal on the left to non-metal on the right.

This is reflected in the graph of melting points for the first 20 elements:

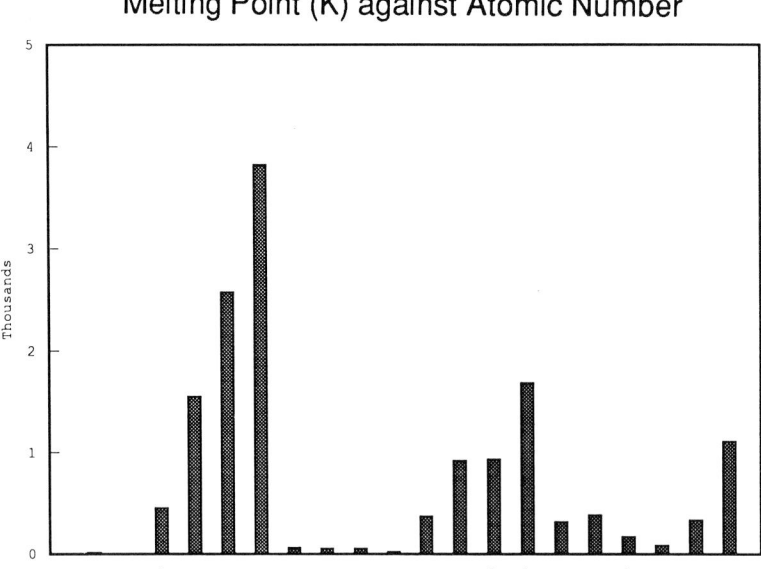

For each Period the melting point rises from Group I to Group IV, then falls to the lowest value at Group VIII. If the d-block elements are also included periodicity can be seen between rows of these elements, but as periodicity becomes less apparent with increasing atomic number this is less obvious than for the s- and p-block elements.

Variation of first ionisation energy with atomic number also shows striking periodicity. The relative position of each Group in relation to the others follows the same pattern in each Period.

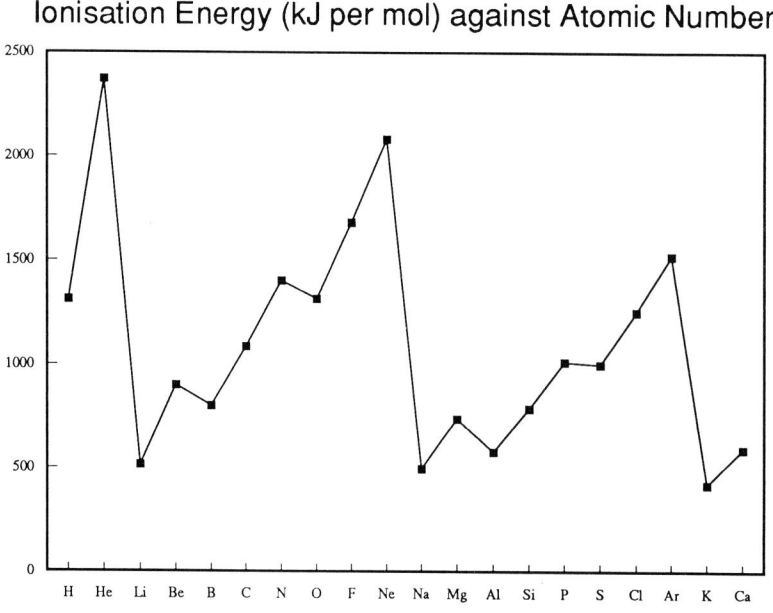

Periodicity is also seen for atomic radius and can be summarised by indicating the main trends:

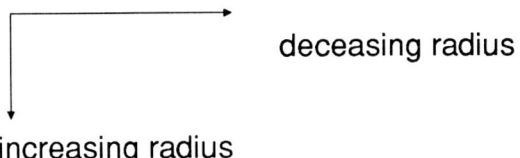

deceasing radius

increasing radius

Some chemical properties of the elements also follow trends and can be summarised in the same way. These include bonding, oxidising properties, acid-base properties of the oxides and electronegativity.

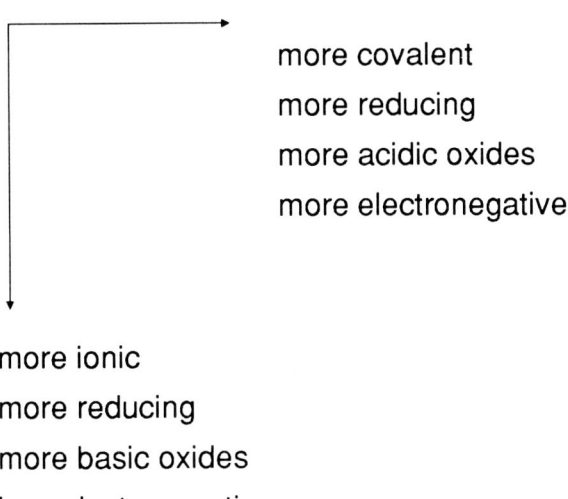

more covalent
more reducing
more acidic oxides
more electronegative

more ionic
more reducing
more basic oxides
less electronegative

Hydrogen

Hydrogen is known by the symbol H and has an electron configuration $1s^1$.

Appearance
Hydrogen is a colourless, odourless, tasteless gas.

General Reactivity
Hydrogen forms more compounds than any other element. The great majority of these compounds are covalent, but the cation H^+ is also very important chemically because of its role in acid-base reactions. Hydrogen is also a powerful reducing agent.

Occurrence and Extraction
Hydrogen is the most abundant element in the universe. There is very little free hydrogen in the earth's atmosphere, but large quantities are found in the combined state as water and organic compounds. Most hydrogen is manufactured from natural gas, which is composed largely of methane.

Physical Properties
Hydrogen is a diatomic gas which has the lowest density of all gases at room temperature and pressure. It is flammable. The splint test is used in the laboratory as a quick test for hydrogen, as this gas gives a mild explosive reaction in the presence of air.

There are 3 isotopes of hydrogen;

protium   - mass number 1
deuterium - mass number 2
tritium   - mass number 3

Chemical Properties
Hydrogen is covalently bonded in almost all its compounds. This is mainly because its ionisation energy is very high, so the formation of H^+ is not favoured. Also, H^+ is a proton and so is extremely small, and this small size gives it exceptionally strong polarising power. Important compounds containing hydrogen are discussed under the other element(s) concerned.

The cation H^+ acts as an extremely strong Lewis acid in water and attaches strongly to a water molecule forming H_3O^+. This ion plays a central role in the mechanism of acid-base reactions.

The most common oxidation number of hydrogen in its compounds is +1, eg: HCl, H₂O, but in compounds containing the hydride ion H⁻ it has an oxidation number of -1.

Industrial Information

The controlled explosive reaction between hydrogen and oxygen is used to power space vehicles.

Hydrogen is also used to reduce nitrogen gas to ammonia in the Haber-Bosch synthesis (see Group V). This is the principal method by which atmospheric nitrogen is brought into the food chain.

Further Information

For further information look up the hydrogen entry.

Data

Ionisation Energy/kJ mol^{-1}   1312

Radius of H$^+$ ion/m 10^{-15}

Group I - *The Alkali Metals*

The elements of Group I, the Alkali metals, are;

	symbol	electron configuration
lithium	Li	[He]2s^1
sodium	Na	[Ne]3s^1
potassium	K	[Ar]4s^1
rubidium	Rb	[Kr]5s^1
caesium	Cs	[Xe]6s^1
francium	Fr	[Rn]7s^1

In each element the valence electron configuration is ns^1, where n is the Period number. The last element, francium, is radioactive and will not be considered here.

Appearance

All the Group I elements are silvery-coloured metals. They are soft, and can be easily cut with a knife to expose a shiny surface which dulls on oxidation.

General Reactivity

These elements are highly reactive metals. The reactivity increases on descending the Group from lithium to caesium. There is a closer similarity between the elements of this Group than in any other Group of the Periodic Table.

Occurrence and Extraction

These elements are too reactive to be found free in nature. Sodium occurs mainly as NaCl (salt) in sea-water and dried-up sea beds. Potassium is more widely distributed in minerals such as sylvite, KCl, but is also extracted from sea-water. The alkali metals are so reactive they cannot be displaced by another element, so are isolated by electrolysis of their molten salts.

Physical Properties

The alkali metals differ from other metals in several ways. They are soft, with low melting and boiling temperatures. They have low densities - Li, Na and K are less dense than water. They have low standard enthalpies of melting and vaporization. They show relatively weak metallic bonding as only one electron is available from each atom.

Alkali metals colour flames. When the element is placed in a flame the heat provides sufficient energy to promote the outermost electron to a higher energy level. On returning to ground level, energy is emitted and this energy has a wavelength in the visible region:

    Li red   Na yellow   K lilac   Rb red   Cs blue

The ionic radii of the alkali metals are all much smaller than the corresponding atomic radii. This is because the atom contains 1 electron in an s level relatively far from the nucleus in a new quantum shell, and when it is removed to form the ion the remaining electrons are in levels closer to the nucleus. In addition, the increased effective nuclear charge attracts the electrons towards the nucleus and decreases the size of the ion.

Chemical Properties

The alkali metals are strong reducing agents. The standard electrode potentials all lie between -2.7V and -3.0V, indicating a strong tendency to form cations in solution. They can reduce oxygen, chlorine, ammonia and hydrogen. The reaction with oxygen tarnishes the metals in air, so they are stored under oil. They cannot be stored under water because they react with it to produce hydrogen and alkali hydroxides;

$2Na(s) + 2H_2O(l) \rightarrow 2NaOH(aq) + H_2(g)$

This reaction illustrates the increasing reactivity on descending the Group. Li dissolves steadily in water with effervescence; sodium reacts more violently and can burn with an orange flame; K ignites on contact with water and burns with a lilac flame; Cs sinks in water, and the rapid generation of hydrogen gas under water produces a shock wave that can shatter a glass container.

Na dissolves in liquid ammonia to give a deep blue solution of sodium cations and solvated electrons. This solution is used as a reducing agent. At higher concentrations the colour of the solution changes to bronze and it conducts electricity like a metal.

The chemistry of Li shows some anomalies, as the cation Li^+ is so small it polarises anions and so introduces a covalent character to its compounds. Li has a diagonal relationship with magnesium.

Oxides

The alkali metals form ionic solid oxides of compostion M_2O when burnt in air. However, Na also forms the peroxide Na_2O_2 as the main product, and K forms the superoxide KO_2, also as the main product.

Hydroxides

Alkali metal hydroxides are white ionic crystalline solids of formula MOH, and are soluble in water. They are all deliquescent except LiOH. The aqueous solutions are all strongly alkaline (hence the name of this Group) and therefore dangerous to handle. They neutralise acids to form salts, eg;

$NaOH(aq) + HCl(aq) \rightarrow NaCl(aq) + H_2O(l)$

Halides

Alkali metal halides are white ionic crystalline solids. They are all soluble in water except LiF, which has a very high lattice enthalpy arising from the strong electrostatic interaction of the small Li$^+$ and F$^-$ ions.

Oxidation States and Ionisation Energies

Alkali metals have oxidation states of 0 and +1. All the common compounds are based on the M$^+$ ion. This is because the first ionisation energy of these elements is low, and the second ionisation energy much higher. The outermost electron is well shielded from the attractive nuclear charge by filled inner electron levels and so is relatively easy to remove. The next electron is much more difficult to remove as it is part of a full level and is also closer to the nucleus.

Ionisation energy decreases down the Group because the outermost electron is progressively further from the nucleus and so is easier to remove.

Industrial Information

Sodium hydroxide, chloride and carbonate are among the most important industrial chemicals associated with this Group. Sodium hydroxide is produced by the electrolysis of saturated brine in a cell with steel cathodes and titanium anodes. Sodium carbonate is made by the Solvay Process, in which soluble sodium chloride is converted into insoluble sodium hydrogencarbonate and filtered off, then heated to produce the carbonate. However, the principal by-product of this process is calcium chloride, and its deposition in rivers causes environmental concern. The Solvay Process is therefore gradually being replaced by the purification of sodium carbonate from minerals.

Further Information

For further information look up the individual elements.

Data

	Atomic Number	Relative Atomic Mass	Melting Point/K	Density/kg m^{-3}
Li	3	6.94	453.7	534
Na	11	22.99	371.0	971
K	19	39.10	336.8	862
Rb	37	85.47	312.2	1532
Cs	55	132.91	301.6	1873

Ionisation Energies/kJ mol⁻¹

	1st	2nd	3rd
Li	513.3	7298.0	11814.8
Na	495.8	4562.4	6912.0
K	418.8	3051.4	4411.0
Rb	403.0	2632.0	3900.0
Cs	375.7	2420.0	3400.0

	Atomic Radius/nm	Ionic Radius/nm	Standard Electrode Potentials/V
Li	0.152	0.068	-3.04
Na	0.185	0.098	-2.71
K	0.227	0.133	-2.92
Rb	0.247	0.148	-2.92
Cs	0.265	0.167	-2.92

Group II - *The Alkaline Earth Metals*

The elements of Group II, the Alkaline Earth Metals, are:

	symbol	electron configuration
beryllium	Be	[He]2s^2
magnesium	Mg	[Ne]3s^2
calcium	Ca	[Ar]4s^2
strontium	Sr	[Kr]5s^2
barium	Ba	[Xe]6s^2
radium	Ra	[Rn]7s^2

The last element, radium, is radioactive and will not be considered here.

Appearance

The Group II elements are all metals with a shiny, silvery-white colour.

General Reactivity

The alkaline earth metals are high in the reactivity series of metals, but not as high as the alkali metals of Group I.

Occurrence and Extraction

These elements are all found in the earth's crust, but not in the elemental form as they are so reactive. Instead, they are widely distributed in rock structures. The main minerals in which magnesium is found are carnellite, magnesite and dolomite. Calcium is found in chalk, limestone, gypsum and anhydrite. Magnesium is the eighth most abundant element in the earth's crust, and calcium is the fifth.

Only magnesium of the elements in this Group is produced on a large scale. It is extracted from sea-water by the addition of calcium hydroxide, which precipitates out the less soluble magnesium hydroxide. This hydroxide is then converted to the chloride, which is electrolysed in a Downs cell to extract magnesium metal.

Physical Properties

The metals of Group II are harder and denser than sodium and potassium, and have higher melting points. These properties are due largely to the presence of 2 valence electrons on each atom, which leads to stronger metallic bonding than occurs in Group I.

Three of these elements give characteristic colours when heated in a flame:

Mg   brilliant white   Ca   brick-red   Sr   crimson   Ba   apple green

Atomic and ionic radii increase smoothly down the Group. The ionic radii are all much smaller than the corresponding atomic radii. This is because the atom contains 2 electrons in an s level relatively far from the nucleus, and it is these electrons which are removed to form the ion. Remaining electrons are thus in levels closer to the nucleus, and in addition the increased effective nuclear charge attracts the electrons towards the nucleus and decreases the size of the ion.

Chemical Properties

The chemical properties of Group II elements are dominated by the strong reducing power of the metals. The elements become increasingly electropositive on descending the Group.

Once started, the reactions with oxygen and chlorine are vigorous:

$2Mg(s) + O_2(g) \rightarrow 2MgO(s)$

$Ca(s) + Cl_2(g) \rightarrow CaCl_2(s)$

All the metals except beryllium form oxides in air at room temperature which dulls the surface of the metal. Beryllium is so reactive it is stored under oil.

All the metals except beryllium reduce water and dilute acids to hydrogen:

$Mg(s) + 2H^+(aq) \rightarrow Mg(aq) + H_2(g)$

Magnesium reacts only slowly with water unless the water is boiling, but calcium reacts rapidly even at room temperature, and forms a cloudy white suspension of sparingly soluble calcium hydroxide.

Calcium, strontium and barium can reduce hydrogen gas when heated, forming the hydride:

$Ca(s) + H_2(g) \rightarrow CaH_2(s)$

The hot metals are also sufficiently strong reducing agents to reduce nitrogen gas and form nitrides:

$3Mg(s) + N_2(g) \rightarrow Mg_3N_2(s)$

Magnesium can reduce, and burn in, carbon dioxide:

$2Mg(s) + CO_2(g) \rightarrow 2MgO(s) + C(s)$

This means that magnesium fires cannot be extinguished using carbon dioxide fire extinguishers.

Oxides

The oxides of alkaline earth metals have the general formula MO and are basic. They are normally prepared by heating the hydroxide or carbonate to release carbon dioxide gas. They have high lattice enthalpies and melting points. Peroxides, MO_2, are known for all these elements except beryllium, as the Be^{2+} cation is too small to accommodate the peroxide anion.

Hydroxides

Calcium, strontium and barium oxides react with water to form hydroxides;

$CaO(s) + H_2O(l) \rightarrow Ca(OH)_2(s)$

Calcium hydroxide is known as slaked lime. It is sparingly soluble in water and the resulting mildly alkaline solution is known as lime water which is used to test for the acidic gas carbon dioxide.

Halides

The Group II halides are normally found in the hydrated form. They are all ionic except beryllium chloride. Anhydrous calcium chloride has such a strong affinity for water it is used as a drying agent.

Oxidation States and Ionisation Energies

In all their compounds these metals have an oxidation number of +2 and, with few exceptions, their compounds are ionic. The reason for this can be seen by examination of the electron configuration, which always has 2 electrons in an outer quantum level. These electrons are relatively easy to remove, but removing the third electron is much more difficult, as it is close to the nucleus and in a filled quantum shell. This results in the formation of M^{2+}. The ionisation energies reflect this electron arrangement. The first two ionisation energies are relatively low, and the third very much higher.

Industrial Information

Magnesium is the only Group II element used on a large scale. It is used in flares, tracer bullets and incendiary bombs as it burns with a brilliant white light. It is also alloyed with aluminium to produce a low-density, strong material used in aircraft. Magnesium oxide has such a high melting point it is used to line furnaces.

Further Information

For further information look up the individual elements.

Data

	Atomic Number	Relative Atomic Mass	Melting Point/K	Density/kg m^{-3}
Be	4	9.012	1551	1847.7
Mg	12	24.31	922	1738
Ca	20	40.08	1112	1550
Sr	38	87.62	1042	2540
Ba	56	137.33	1002	3594

Ionisation Energies/kJ mol^{-1}

	1st	2nd	3rd
Be	899.4	1757.1	14848
Mg	737.7	1450.7	7732.6
Ca	589.7	1145	4910
Sr	549.5	1064.2	4210
Ba	502.8	965.1	3600

	Atomic Radius/pm	Ionic Radius/nm (M^{2+})	Standard Electrode Potentials/V
Be	113.3	0.034	-1.85
Mg	160	0.078	-2.36
Ca	197.3	0.106	-2.87
Sr	215.1	0.127	-2.89
Ba	217.3	0.143	-2.90

Group III

The elements of Group III are:

	symbol	electron configuration
boron	B	$[He]2s^2 2p^1$
aluminium	Al	$[Ne]3s^2 3p^1$
gallium	Ga	$[Ar]3d^{10} 4s^2 4p^1$
indium	In	$[Kr]4d^{10} 5s^2 5p^1$
thallium	Tl	$[Xe]4f^{14} 5d^{10} 6s^2 6p^1$

Appearance

Boron is a non-metallic grey powder, and all the other members of Group III are soft, silvery metals. Thallium develops a bluish tinge on oxidation.

General Reactivity

The general trend down Group III is from non-metallic to metallic character. Boron is a non-metal with a covalent network structure. The other elements are considerably larger than boron and consequently are more ionic and metallic in character. Aluminium has a close-packed metallic structure but is on the borderline between ionic and covalent character in its compounds. The remainder of Group III are generally considered to be metals, although some compounds exhibit covalent characteristics.

Occurrence and Extraction

These elements are not found free in nature, but are all present in various minerals and ores. The most important aluminium-containing minerals are bauxite and cryolite.

Aluminium is the most widely used element in this Group. It is obtained by the electrolysis of aluminium oxide, which is purified from bauxite. The melting point of the aluminium oxide is too high for electrolysis of the melt, so instead it is dissolved in molten cryolite.

Physical Properties

The influence of the non-metallic character in this Group is reflected by the softness of the metals. The melting points of all the elements are high, but the melting point of boron is much higher than that of beryllium in Group II, whereas the melting point of aluminium is similar to that of magnesium in Group II. The densities of all the Group III elements are higher than those of Group II elements.

The ionic radii are much smaller than the atomic radii. This is because the atom contains 3 electrons in a quantum level relatively far from the nucleus, and when they are removed to form the ion the remaining electrons are in levels closer to the nucleus. In addition, the increased effective nuclear charge attracts the electrons towards the nucleus and decreases the size of the ion.

Chemical Properties

The chemical properties of Group III elements reflect the increasingly metallic character of descending members of the Group. Here only boron and aluminium will be considered.

Boron is chemically unreactive except at high temperatures. Aluminium is a highly reactive metal which is readily oxidised in air. This oxide coating is resistant to acids but is moderately soluble in alkalis. Aluminium can therefore reduce strong alkalis, a product being the tetrahydroxoaluminate ion, $Al(OH)_4^-$. Aluminium also reacts violently with iron (III) oxide to produce iron in the Thermit process:

$2Al(s) + Fe_2O_3 \rightarrow 2Fe(s) + Al_2O_3(s)$

Oxides

Boron oxide, B_2O_3, is an insoluble white solid with a very high boiling point (over 2000K) because of its extended covalently-bonded network structure. Aluminium oxide, Al_2O_3, is amphoteric.

Halides

The most important halide of boron is boron trifluoride, which is a gas.

Aluminium chloride, $AlCl_3$, is a volatile solid which sublimes at 458K. The vapour formed on sublimation consists of an equilibrium mixture of monomers ($AlCl_3$) and dimers (Al_2Cl_6). It is used to prepare the powerful and versatile reducing agent lithium tetrahydridoaluminate, $LiAlH_4$.

Both boron chloride and aluminium chloride act as Lewis acids to a wide range of electron-pair donors, and this has led to their widespread use as catalysts. Aluminium chloride is used in the important Friedel-Crafts reaction.

Hydrides

Boron forms an extensive and interesting series of hydrides, the boranes. The simplest of these is not BH_3 as expected, but its dimer B_2H_6.

Oxidation States and Ionisation Energies

Boron and aluminium occur only with oxidation number +3 in their compounds, and with a few exceptions their compounds are best described as ionic. The electron configuration shows 3 electrons outside a noble gas configuration, 2 in an s shell and 1 in a p shell. The outermost p electron is easy to remove as it is furthest from the nucleus and well shielded from the effective nuclear charge. The next 2 s electrons are also relatively easy to remove. Removal of any further electrons disturbs a filled quantum shell so is difficult. This is reflected in the ionisation energies. The first 3 ionisation energies are low, and the fourth very much higher.

Industrial Information

Boron has limited uses, but is used in flares to provide a highly visible green colour. Boron filaments are now used extensively in the aerospace industry as a light weight yet strong material. Boracic acid is used as a mild antiseptic, and borax as a water softener in washing powders. Borosilicate glass contains boron compounds.

Aluminium is one of the most industrially important materials. It is light, non-toxic, has a high thermal conductivity, can be easily worked and does not corrode due to its oxide coating, which is very effective although only 10nm thick. It has several domestic uses such as cooking utensils, aluminium foil and bottle tops, and is widely used in the building industry where a strong, light, easily-constructed material is required. These properties also make it invaluable in the building of aeroplanes and spacecraft.

Further Information

For further information look up the individual elements.

Data

	Atomic Number	Relative Atomic Mass	Melting Point/K	Density/kg m^{-3}
B	5	10.81	2573	2340
Al	13	26.98	933.52	2698
Ga	31	69.72	302.9	5907
In	49	114.82	429.32	7310
Tl	81	204.38	576.7	11850

Ionisation Energies/kJ mol^{-1}

	1st	2nd	3rd	4th
B	800.6	2427	3660	25025
Al	577.4	1816.6	2744.6	11575
Ga	578.8	1979	2963	6200
In	558.3	1820.6	2704	5200
Tl	589.3	1971	2878	4900

	Atomic Radius/nm	Ionic Radius/nm (M^{3+})
B	0.079	
Al	0.1431	0.057
Ga	0.1221	0.062
In	0.1626	0.092
Tl	0.1704	0.105

Group IV

The elements of Group IV are:

	symbol	electron configuration
carbon	C	[He]$2s^2 2p^2$
silicon	Si	[Ne]$3s^2 3p^2$
germanium	Ge	[Ar]$3d^{10} 4s^2 4p^2$
tin	Sn	[Kr]$4d^{10} 5s^2 5p^2$
lead	Pb	[Xe]$4f^{14} 5d^{10} 6s^2 6p^2$

Appearance

The expected similarity in appearance between elements in the same Group is much less apparent in Group IV, where there is a considerable change in character on descending the Group. Carbon is a dull black colour in the form of graphite, or hard and transparent in the form of diamond; silicon and germanium are dull grey or black; tin and lead are a shiny grey colour.

General Reactivity

In Group IV the elements change from non-metallic in character at the top of the Group to metallic at the bottom. Carbon is a non-metal, silicon and germanium are metalloids, and tin and lead are typical metals. The general reactivity of the Group as a whole is therefore difficult to ascertain, and the reactivity of each element must be considered individually.

Carbon exists in two important allotropic forms, diamond and graphite. Diamond has an extended covalently-bonded structure in which each carbon atom is bonded to 4 others. This compact, rigid arrangement explains why diamond is both extremely hard and chemically inert. Graphite has a layer structure. Planes of covalently-bonded carbon atoms are held together by weak van der Waals forces, and slide over each other easily. Chemically, graphite is more reactive than diamond but still does not react easily. However, it does oxidise at high temperatures and this is the reason why carbon is used in various forms as fuel.

Silicon is chemically unreactive.

Germanium is also unreactive and not widely used, so will not be considered further. It does, however, have excellent semi-conducting properties so may become more widely used in a few years' time.

Both tin and lead are generally unreactive metals. Tin has 2 common allotropes. At room temperature the stable form is white tin; below 286.2K the stable form is grey tin.

Tin has a tendency to displace lead, and not vice versa as may be expected.

Occurrence and Extraction

Carbon, tin and lead can all be found in the elemental form in the earth's crust, and are readily mined.

Silicon is found in mineral deposits and purified from them. Very pure silicon is required for semi-conductors, and is obtained from silicon (IV) chloride. This is first purified by fractional distillation, then reduced to give the element. The silicon is then further purified by zone refining, in which a molten zone is moved along a silicon rod several times, carrying impurities to one end where they are removed.

Physical Properties

The physical properties of Group IV elements vary quite widely from one element to another, consistent with the increasing metallic character on descending the Group. The structures change from giant molecular lattices in carbon and silicon to giant metallic lattices in tin and lead, and this is the reason for the changes in physical properties. The change in bonding from covalent to metallic down the Group causes a decrease in melting point, boiling point, heat of atomisation and first ionisation energy. At the same time, the increasing metallic character causes a general increase in density and conductivity. Carbon and silicon do not conduct electricity.

Diamond has a very high refractive index (the reason for its sparkle) and this, along with its rarity, has made it valuable as a jewel. However, it is also the hardest natural substance known and so is important industrially.

The most important physical property of silicon is that it is a semi-conductor. Small silicon chips, just a few millimetres square, have revolutionised the computer and microprocessor industries.

Tin and lead, as typical metals, are good conductors of electricity.

Chemical Properties

In general, chemical reactivity increases on descending the Group.

The first member of the Group, carbon, is strikingly different from the others as it has the unique ability to form stable compounds containing long chains and rings of carbon atoms. This property, called catenation, results in carbon forming an enormous range of organic compounds. The ability to catenate results from the C-C bond having almost the same bond energy as the C-O bond, so that oxidation of carbon compounds is energetically favourable. Also, the small size of the carbon atom allows 2 carbon atoms to approach close together and allow overlap of p-orbitals, so that multiple bonds can be formed. The organic compounds formed from carbon have a chemistry entirely different to any inorganic counterpart.

The C-C and C-Si bond energies are very similar, so many organo-silicon compounds are known. Silicon does not, however, form multiple bonds.

Silicon is unreactive chemically because an oxide layer seals the surface from attack, and high temperatures are required for oxidation to occur. Silicon does, however, react with fluorine at room temperature. It is not attacked by aqueous acids, but does react with concentrated alkalis.

Tin and lead are quite easily oxidised, tin usually to tin (IV) and lead to lead (II). Both tin and lead reduce the halogens.

Oxides

There are many oxides of Group IV elements. The major oxides are:

CO (gas), CO_2 (gas), SiO_2 (solid), SnO (solid), SnO_2 (solid), PbO (solid), Pb_3O_4 (solid), PbO_2 (solid).

Oxides with a lower oxidation number become more stable going down the Group.

Carbon dioxide is essential to life as it is the source of carbon for plants. During photosynthesis carbon is combined with water to form carbohydrates. Solid carbon dioxide sublimes directly to a gas, so is widely used as a cheap refrigerant.

Silica forms a strong bond with oxygen in silica, SiO_2, one form of which is quartz. Sand is impure quartz. Oxoanions derived from silica are called silicates. They are very common in nature and have structures based on SiO_4 tetrahedra. One of these is asbestos.

Tin (IV) oxide, tin (II) oxide and lead (II) oxide are amphoteric.

Halides

All the elements of Group IV form tetrahalides, but only tin and lead form dihalides. The tetrahalides are covalent tetrahedral molecules whereas the dihalides are best regarded as ionic.

Hydrides

The hydrides of carbon are the hydrocarbons - organic compounds.

Silicon forms a series of hydrides called the silanes, with the general formula Si_nH_{2n+2}.

Oxidation States and Ionisation Energies

Group IV elements exist in 2 oxidation states, +2 and +4. There is a steady increase in the stability of the +2 oxidation state on descending the Group.

The elements in this Group have 4 electrons in their outermost shell, 2 s electrons and 2 p electrons. The first 4 ionisation energies rise in a fairly even manner, and the 5th ionisation energy is very large, reflecting the removal of an electron from a filled level nearer to the nucleus. Compounds of tin and lead in which the Group IV element has an oxidation number +2 are regarded as ionic. In these compounds, the Sn^{2+} and Pb^{2+} ions are formed by loss of the outermost 2 electrons, whilst the 2 s electrons remain relatively stable in their filled sub-shell. This is called the "inert pair effect".

Industrial Information

The industrial importance of carbon in petrochemicals is immense. These hydrocarbons are used extensively in almost all areas of modern civilisation; clothing, dyes, fertilisers, agrochemicals, fuels and new materials.

Silicon also contributes to new technology in the silicon chip, which has revolutionised the computer and high-tech industries. Germanium is, like silicon, a semi-conductor used in similar devices, but it is less widely used.

Tin and lead have more traditional industrial uses. Tin is used as a coating for other metals to prevent corrosion, such as in tin cans, but it is in alloys that tin is used most extensively. These alloys include bronze, soft solder, type metal, phosphor bronze and pewter. Lead is used in great quantities in storage batteries. It is also used in cable covering, ammunition and in the manufacture of tetraethyl lead, an anti-knocking compound added to petrol.

Further Information

For further information look up the individual elements.

Data

	Atomic Number	Relative Atomic Mass	Melting Point/K	Density/kg m^{-3}
C	6	12.011	3820 (diamond)	3513
Si	14	28.086	1683	2329
Ge	32	72.61	1210.6	5323
Sn	50	118.71	505.1	7130
Pb	82	207.2	600.65	11350

Ionisation Energies/kJ mol^{-1}

	1st	2nd	3rd	4th	5th
C	1086.2	2352	4620	6222	37827
Si	786.5	1577.1	3231.4	4355.5	16091
Ge	762.1	1537	3302	4410	9020
Sn	708.6	1411.8	2943	3930.2	6974
Pb	715.5	1450.4	3081.5	4083	6640

	Atomic Radius/nm	Ionic Radii/nm
C	0.077	
Si	0.117	
Ge	0.122	
Sn	0.140	0.093 (Sn^{2+}) 0.074 (Sn^{4+})
Pb	0.154	0.132 (Pb^{2+}) 0.084 (Pb^{4+})

Group V

The elements of Group V are:

	symbol	electron configuration
nitrogen	N	$[He]2s^2 2p^3$
phosphorus	P	$[Ne]3s^2 3p^3$
arsenic	As	$[Ar]3d^{10} 4s^2 4p^3$
antimony	Sb	$[Kr]4d^{10} 5s^2 5p^3$
bismuth	Bi	$[Xe]4f^{14} 5d^{10} 6s^2 6p^3$

The most important members of this Group are nitrogen and phosphorus. The other elements will mostly not be considered here.

Appearance

The appearance of the Group V elements varies widely. Nitrogen is a colourless, odourless gas; phosphorus exists in white, red and black solid forms; arsenic is found in yellow and grey solid forms; antimony is found in a metallic or amorphous grey form; and finally bismuth is a white, crystalline, brittle metal. These appearances reflect the changing nature of the elements as the Group is descended, from non-metal to metal.

General Reactivity

The elements of Group V show a marked trend towards metallic character on descending the Group. This trend is reflected both in their structures and in their chemical properties, as for example in the oxides which become increasingly basic.

Occurrence

Nitrogen is found in the atmosphere, and makes up 78% of the air by volume. Phosphorus is not found free in nature, but occurs in several minerals and ores such as phosphate rock. The other elements are all found in the elemental form in the earth's crust, but more frequently as minerals.

Physical Properties

The physical properties of this Group vary widely as nitrogen is a gas, and the other elements are solids of increasingly metallic character.

Nitrogen exists as the diatomic molecule N_2. It is a colourless, odourless gas, which condenses to a colourless liquid at 77K. The strength of the bond and the short bond length provide evidence for the bond between the N atoms being a triple bond.

Phosphorus has at least 2 allotropes, red and white phosphorus. White phosphorus is a solid composed of covalent tetrahedral P_4 molecules, and red phosphorus is an amorphous solid which has an extended covalent structure.

The covalent radii of the atoms increase on descending the Group. However, the N atom is anomalously small and so it can multiple-bond to other N, C and O atoms.

Chemical Properties

Both nitrogen and phosphorus exist in oxidation states +3 and +5 in their compounds. Nitrogen is very unreactive, mainly because its bond enthalpy is very high (944 kJ mol^{-1}). The only element to react with nitrogen at room temperature is lithium, to form the nitride Li_3N. Magnesium also reacts directly, but only when ignited. Some micro-organisms, however, have developed a mechanism for reacting directly with nitrogen gas and building it into protein - this is called nitrogen fixation, and is an important early step in the food chain.

Phosphorus is more reactive than nitrogen. It reacts with metals to form phosphides, with sulphur to form sulphides, with halogens to form halides, and ignites in air to form oxides. It also reacts with both alkalis and concentrated nitric acid.

Oxides

There are 5 oxides of nitrogen, with N ranging in oxidation number from +1 to +5; N_2O, NO, N_2O_3, NO_2, N_2O_5. There are also 2 important oxoacids of nitrogen - nitric (III) acid (nitrous acid) HONO, and nitric (V) acid (nitric acid) HNO_3. Nitric acid is highly reactive, and behaves as an oxidising agent and a nitrating agent as well as an acid.

There are many oxoacids of phosphorus, the most important being phosphoric acid $(HO)_3PO$. This is produced on a large scale commercially as it is used in the manufacture of fertilisers.

Halides

The nitrogen halides $N(hal)_3$ all have covalent, pyramidal structures. More important are the 2 series of phosphorus halides, $P(hal)_3$ and $P(hal)_5$

Compounds with hydrogen

The most important of these is ammonia, NH_3. During the industrial synthesis of ammonia the inert nitrogen obtained from the air is made into a reactive compound, ammonia, thus making atmospheric nitrogen available for many reactions. Ammonia is a reducing agent, but can be reduced by stronger reducing agents such as sodium metal. Ammonia is also a Lewis base, as it has a lone pair of electrons.

Phosphine, PH_3, is also a Lewis base but is less soluble in water than ammonia because it does not form hydrogen bonds.

Oxidation States and Ionisation Energies

The elements of this Group have the general electron configuration ns^2np^3. They all exhibit an expected oxidation state of +3 arising from the 3 unpaired p electrons, but also an oxidation state of +5. In all the elements except N this is made possible by promotion of an s electron to an available d orbital. Nitrogen is remarkable for its wide range of oxidation states ranging from -3 to +5.

In this Group the first 5 ionisation energies are relatively low, reflecting the removal of the 2 s and 3 p electrons. There is a larger increase between the 5th and 6th ionisation energies as an electron is removed from the inner, filled quantum shell.

Industrial Information

For industrial use nitrogen is obtained by fractional distillation of the air. It is used for the manufacture of ammonia by the Haber-Bosch Process:

$N_2(g) + 3H_2(g) \rightarrow 2NH_3(g)$

A catalyst of finely-divided iron is required for this equilibrium reaction. The mixture is heated to 450 °C at 250 atm of pressure. The yield of ammonia is increased by working at high pressure, but this adds to the cost of the plant, and a compromise between cost and yield is needed. The reaction is exothermic so the yield of ammonia is increased by low temperatures, but this slows up the reaction so again a compromise is needed. At the temperatures and pressures used in practice about 15% conversion is attained. The ammonia is condensed and removed from the plant and the unreacted gases are recirculated.

Ammonia has numerous uses - approximately 100 megatonnes are produced worldwide each year. It is used as a fertiliser both directly and after conversion to other fertilisers such as ammonium nitrate. It is also a raw material for nitric acid manufacture and for the production of nylon.

Gaseous nitrogen is used to provide an inert atmosphere for reactions which cannot be carried out in oxygen. It is also used as a carrier gas in liquid-gas chromatography.

Phosphorus is used in match-heads and on safety match boxes.

Further Information

For further information look up the individual elements.

Data

	Atomic Number	Relative Atomic Mass	Melting Point/K	Density/kg m^{-3}
N	7	14.007	63.29	1.2506
P	15	30.974	317.3	1820 (white)
As	33	74.92	1090	5780
Sb	51	121.75	903.89	6691
Bi	83	208.98	544.5	9747

Ionisation Energies/kJ mol^{-1}

	1st	2nd	3rd	4th	5th	6th
N	1402.3	2856.1	4578	7474.9	9440	53265.6
P	1011.7	1903.2	2912	4956	6273	21268
As	947.0	1798	2735	4837	6042	12305
Sb	833.7	1794	2443	4260	5400	10400
Bi	703.2	1610	2466	4372	5400	8520

	Atomic Radius/nm	Covalent Radius/nm	Ionic Radius/nm
N	0.0549	0.070	
P	0.1105	0.110	
As	0.1245	0.121	
Sb	0.145	0.141	
Bi	0.154	0.146	0.117(Bi^{3+})

Group VI

The elements of Group VI are:

	symbol	electron configuration
oxygen	O	$[He]2s^2 2p^4$
sulphur	S	$[Ne]3s^2 3p^4$
selenium	Se	$[Ar]3d^{10} 4s^2 4p^4$
tellurium	Te	$[Kr]4d^{10} 5s^2 5p^4$
polonium	Po	$[Xe]4f^{14} 5d^{10} 6s^2 6p^4$

Appearance

The first element of this Group, oxygen, is the only gas, and is colourless and odourless. Sulphur is a pale yellow, brittle solid. Selenium can have either an amorphous or a crystalline structure; the amorphous form can be red or black, and the crystalline form can be red or grey. Tellurium is a silvery-white colour with a metallic lustre. Polonium is a naturally radioactive element.

Selenium and tellurium are rare elements with few uses, and along with polonium will not be considered further here.

General Reactivity

Oxygen and sulphur are highly electronegative elements - the electronegativity of oxygen is second only to that of fluorine. Their general reactivity is therefore dominated by their ability to gain electrons.

There is a transition down the Group from non-metallic to more metallic properties, so that oxygen is a non-metal and tellurium a metalloid. All the elements except polonium form M^{2-} ions.

There is a marked difference between oxygen and the other members of the Group. This arises from

(a) the small size of the O atom which enables it to form multiple bonds

(b) its inability to expand its valence shell like the other elements as it has no accessible d-orbitals

(c) its high electronegativity, which enables it to participate in hydrogen-bonding.

Occurrence and Extraction

Oxygen occurs widely as the free element in the form of O_2, comprising 21% of the air by volume. It also occurs as O_3, ozone, at high altitudes in the ozone layer. In the combined form it is found in very many minerals, and also in water. Oxygen is obtained industrially by the fractional distillation of liquid air. It is stored under pressure in cylinders.

Sulphur is found as the free element and also as metal sulphide ores and a number of sulphates. Native sulphur is brought to the surface from underground deposits by the Frasch Process, which uses superheated water to melt the sulphur and force it upwards.

Physical Properties

The covalent and ionic radii increase going down the Group, as electrons occupy shells with higher quantum numbers.

Oxygen occurs as 2 gaseous allotropes, O_2 (dioxygen or more commonly oxygen) and O_3 (trioxygen or ozone). Oxygen is the more common. It condenses to a pale blue liquid at -183°C which is paramagnetic. Ozone is a pale blue, pungent gas which condenses to an inky-blue liquid at -112°C. The ozone layer in the upper atmosphere is an important shield against harmful ultra-violet radiation from the sun.

Sulphur has several allotropes, the 2 main ones being rhombic and monoclinic sulphur. These both consist of S_8 molecules.

Chemical Properties

Oxygen is a very reactive oxidising agent, principally in combustion and respiration reactions. Ozone is also a highly reactive and powerful oxidising agent which can cleave the C=C double bond.

Sulphur is reactive in all its forms. It burns in oxygen with a blue flame to form sulphur dioxide, SO_2, a pungent, choking gas. With elements of lower reactivity it acts as an oxidising agent and forms sulphides - this reaction can be vigorous with some metals, especially if the metal is finely divided. Sulphur is not as strong an oxidising agent as oxygen.

Oxides

The most important oxides of sulphur are sulphur dioxide, SO_2, and sulphur trioxide, SO_3. SO_2 forms when sulphur is burnt in air or oxygen, and as all fossil fuels contain sulphur it is formed when they burn and contributes to the problem of acid rain. It is a colourless, toxic, pungent gas and dissloves in water to give sulphurous acid, H_2SO_3. The salts of this acid contain the sulphite ion, SO_3^{2-}. This ion has an important oxidising reaction to the thiosulphate ion, $S_2O_3^{2-}$, which is used for the titrimetric determination of iodine.

Sulphur trioxide is a volatile white solid that reacts violently with water.

Pure sulphuric acid, H_2SO_4, is a colourless, viscous liquid. It is a chemically important reagent as it behaves as an acid, an oxidising reagent and a dehydrating agent. It is also cheap, so is widely used in industry.

Halides

The only halide of oxygen is oxygen difluoride, OF_2, which is a colourless, toxic gas.

Sulphur has numerous halides, the most important being sulphur hexafluoride, SF_6, and disulphur dichloride, S_2Cl_2.

Compounds with hydrogen

The most important of these is water, H_2O, one of the most versatile of chemicals. It can act as a Bronsted acid or base, a Lewis base, an oxidising agent and a reducing agent.

Hydrogen peroxide, H_2O_2, is a pale blue liquid resembling water in its physical properties as both are extensively hydrogen-bonded. It has a strong oxidising ability and this makes it useful industrially.

Hydrogen sulphide, H_2S, is commonly known as "bad egg gas" because of its smell. It dissolves readily to form a weakly acidic solution, and is a strong reducing agent.

Oxidation States and Electron Affinities

The oxidation number of oxygen in its compounds is almost always -2. The oxidation numbers of sulphur range from -2 to +6, but the most common are -2, +4 and +6. This wide range is partly due to sulphur's ability to accommodate extra electrons in its valence shell by using available d-orbitals.

The 1st electron affinity (electron gain) is exothermic, but the 2nd is strongly endothermic and so overall the formation of O^{2-} is endothermic. This is usually compensated by a high lattice enthalpy. Remember that electron affinities are quoted as $-E$ kJ mol^{-1}.

Industrial Information

Sulphuric acid is of immense industrial importance. Because it has 3 chemical functions and is very cheap to produce, sulphuric acid is used at some stage of the manufacture of most products. It is said that the economic prosperity of a country can be assessed by its consumption of sulphuric acid. It is manufactured by the Contact Process.

Hydrogen peroxide is used to bleach hair and textiles, as a mild disinfectant and in pollution control.

Sulphur hexafluoride can be ionized by electric fields and so is widely used as a gaseous insulator in transformers and electrical switch gear.

Further Information

For further information look up the individual elements.

Data

	Atomic Number	Relative Atomic Mass	Melting Point/K	Density/kg m^{-3}
O	8	15.9994	54.8	1.429
S	16	32.066	386	2070
Se	34	78.96	490	4790
Te	52	127.60	722.7	6240

Electron Affinity/kJ mol^{-1}

O → O$^-$	+141
O$^-$ → O^{2-}	-703
S → S$^-$	+200.4
S$^-$ → S^{2-}	-694

Ionisation Energies/kJ mol^{-1}

	1st	2nd	3rd	4th	5th	6th	7th
O	1313.9	3388.2	5300.3	7469.1	10989.3	13326.2	71333
S	999.6	2251	3361	4564	7013	8495	27106
Se	940.9	2044	2974	4144	6590	7883	14990
Te	869.2	1795	2698	3610	5668	6822	13200

	Atomic Radius/nm	Covalent Radius/nm	Ionic Radius/nm (M^{2-})
O	0.0604	0.066	0.138
S	0.1035	0.104	0.184
Se	0.116	0.117	0.198
Te	0.1432	0.137	0.221

Group VII *The Halogens*

The elements of Group VII, the Halogens, are:

	symbol	electron configuration
fluorine	F	[He]$2s^2 2p^5$
chlorine	Cl	[Ne]$3s^2 3p^5$
bromine	Br	[Ar]$3d^{10} 4s^2 4p^5$
iodine	I	[Kr]$4d^{10} 5s^2 5p^5$
astatine	As	[Xe]$4f^{14} 5d^{10} 6s^2 6p^5$

All the isotopes of astatine are radioactive, and so this element will not be considered further here.

Appearance

Fluorine is a poisonous pale yellow gas, chlorine is a poisonous pale green gas, bromine is a toxic and caustic brown volatile liquid, and iodine is a shiny black solid which easily sublimes to form a violet vapour on heating.

General Reactivity

The elements of Group VII, the Halogens, are a very similar set of non-metals. They all exist as diatomic molecules, X$_2$, and oxidise metals to form halides. The halogen oxides are acidic, and the hydrides are covalent. Fluorine is the most electronegative element of all. Generally, electronegativity and oxidising ability decrease on descending the Group. The result of this decreasing electronegativity is increased covalent character in the compounds, so that AlF$_3$ is ionic whereas AlCl$_3$ is covalent.

Fluorine shows some anomalies because of the small size of its atom and ion. This allows several F atoms to pack around a different central atom, as in AlF$_6^{3-}$ compared with AlCl$_4^-$. The F-F bond is also unexpectedly weak because the small size of the F atom brings the lone pairs closer together than in other halogens, and repulsion weakens the bond.

Occurrence and Extraction

The halogens are too reactive to occur free in nature. Fluorine is mined as fluorspar, calcium fluoride and cryolite. It is extracted by electrolysis as no oxidant will oxidise fluorides to fluorine. Chlorine is also found in minerals such as rock-salt, and huge quantities of chloride ions occur in seawater, inland lakes and subterranean brine wells. It is obtained by the electrolysis of molten sodium chloride or brine. Bromine is also found as the bromide ion in seawater, and in larger quantities in brine wells, from which it is extracted. Iodine is mined as sodium iodate(V), NaIO$_3$, which is present in Chile saltpetre. It is obtained by reaction with sodium hydrogensulphite.

Physical Properties

At room temperature all the halogens exist as diatomic molecules. The melting points, boiling points, atomic radii and ionic radii all increase on descending the Group. The shapes of the covalent molecules and ions are readily explained by VSEPR (valence shell electron pair repulsion) theory and these compounds are often used to illustrate the theory. Fluorine is never surrounded by more than 8 electrons, whereas the other halogens may be surrounded by up to 14 electrons.

Chemical Properties

The most characteristic chemical feature of the halogens is their ability to oxidise. Fluorine has the strongest oxidising ability, so other elements which combine with fluorine have their highest possible oxidation number. Fluorine is such a strong oxidising agent that it must be prepared by electrolysis. Chlorine is the next strongest oxidising agent, but it can be prepared by chemical oxidation. Most elements react directly with chlorine, bromine and iodine, with decreasing reactivity going down the Group, but often the reaction must be activated by heat or UV light. The oxidation of thiosulphate ions, $S_2O_3^{2-}$, by the halogens is quantitative. This means that oxidising agents can be estimated accurately; the oxidising agent is reacted with excess I$^-$ ions, and the liberated I$_2$ titrated with standard thiosulphate solution. The end point is detected with starch as indicator, which forms a dark blue complex with iodine. Chlorine, bromine and iodine disproportionate in the presence of water and alkalis.

Oxides and Oxoacids

There are no fluorine oxides as F is more electronegative than O. Chlorine, bromine and iodine each form several oxides which are thermally unstable, such as chlorine dioxide ClO$_2$. The only fluorine oxoacid, HOF, is unstable at room temperature, but there are many oxoacids of the other halogens. The best known salts of these are; hypochlorite, chlorate(I) ClO$^-$; chlorite, chlorate(III) ClO$_2^-$; hypochlorate, chlorate(V) ClO$_3^-$; perchlorate, chlorate(VII) ClO$_4^-$. These are all powerful oxidising agents.

Halides

The halogens can combine with each other to form interhalogens and polyhalide ions.

Polyhalide ions have the general formula [Y-X-Y]$^-$. It is not possible for F to represent X in a polyhalide ion as it cannot expand its octet.

Hydrides

Hydrogen halides have the general formula HX. HF is a colourless liquid which boils at 19.5°C, and all the other hydrogen halides are colourless gases. HF is a liquid due to the extensive hydrogen bonding which occurs between molecules. All the hydrogen halides dissolve easily to give acidic solutions, the most widely used being hydrochloric acid, HCl. All except HF are typical acids; they liberate carbon dioxide from carbonates and form salts with basic oxides. HF is a weak acid because the H-F bond is very strong, and because hydrogen-bonding occurs between F⁻ and HF in solution.

Organic Compounds

The halogens form organic compounds which are best known for their industrial and environmental impact, such as PVC, DDT and TCP.

Oxidation States and Electron Affinities

Fluorine in all its compounds has an assigned oxidation number of -1, as it is the most electronegative of all the elements. The other halogens show a wide range of oxidation numbers, and the redox chemistry of these halogens is important. The oxidation numbers most commonly shown are odd; there are few compounds with even oxidation numbers and they are often thermally unstable. Chlorine is the 3rd most electronegative element after F and O. The halide ions are readily formed by accepting 1 electron, as this completes an octet of valence electrons. The electron affinity decreases on descending the Group.

Industrial information

The halogens are probably the most important Group of the Periodic Table used in industry. Fluorine is widely used as an oxidising agent. HF is used to etch glass. Chlorine is used for chlorinating drinking water, and in many organochlorine compounds. Some of these, such as the insecticide DDT, are effective but environmentally damaging, and much controversy surrounds their use. Chlorine dioxide is used to bleach wood pulp for paper making, as it gives a good whiteness without degrading the paper. Hypochlorites are used in domestic bleaches. Potassium chlorate (V) is used as an oxidant in fireworks and matches. Ammonium chlorate (VII) is used as a fuel in space rockets when mixed with powered aluminium.

Further Information

For further information look up the individual elements.

Data

	Atomic Number	Relative Atomic Mass	Boiling Point/K	Density/kg m⁻³
F	9	19.99	885.01	1.696
Cl	17	35.45	3239.18	3.214
Br	35	79.90	4331.93	7.59(gas)
I	53	126.90	4457.50	4930

Electron Affinity(M-M⁻)kJ mol^{-1}

F	333
Cl	348
Br	324
I	295

Ionisation Energies/kJ mol^{-1}

	1st	2nd	3rd	4th
F	1681	3374	6050	8408
Cl	1251.1	2297	3826	5158
Br	1139.9	2104	3500	4560
I	1008.4	1845.9	3200	4100

	5th	6th	7th	8th
F	11023	15764	17867	92036
Cl	6540	9362	11020	33610
Br	5760	8550	9940	18600
I	5000	7400	8700	16400

	Atomic Radius/nm	Covalent Radius/nm	Ionic Radius/nm (X⁻)
F	0.0709	0.064	0.133
Cl	0.0994	0.099	0.181
Br	0.1145	0.1142	0.196
I	0.1331	0.1333	0.220

Group VIII - *The Noble Gases*

The elements of Group VIII, the noble gases, are:

	symbol	electron configuration
helium	He	$1s^2$
neon	Ne	$[He]2s^2 2p^6$
argon	Ar	$[Ne]3s^2 3p^6$
krypton	Kr	$[Ar]3d^{10} 4s^2 4p^6$
xenon	Xe	$[Kr]4d^{10} 5s^2 5p^6$
radon	Rn	$[Xe]4f^{14} 5d^{10} 6s^2 6p^6$

Radon is a hazardous radioactive gas and will not be considered further here.

Appearance

As the name suggests, all the elements in this Group are gases.

General Reactivity

These elements are generally considered unreactive, because they have closed-shell configurations.

Occurrence and Extraction

The noble gases are all found in minute quantities in the atmosphere, and are isolated by fractional distillation of liquid air. Helium can be obtained from natural gas wells where it has accumulated as a result of radioactive decay.

Physical Properties

All these gases are monatomic. They boil at low temperatures as only dispersion forces act between the atoms. Helium has the lowest boiling point of any substance at 4.2K.

Atomic radii increase on descending the Group.

Chemical Properties

This Group was originally named the "inert gases", as it was thought they formed no compounds. However, compounds of these gases are now well documented. Helium, neon and argon form no known compounds.

Krypton forms KrF_2, a colourless solid, on reaction with fluorine.

Xenon forms a wide range of compounds with oxygen and fluorine.

Oxidation States and Ionisation Energies

The 1st ionisation energy decreases on descending the Group, as the valence shell becomes further away from the nucleus and electrons easier to remove. The first ionisation energy of xenon is comparable with that of bromine, which explains why xenon forms compounds with oxygen and fluorine relatively easily. The oxidation numbers of xenon in its compounds are +2, +4, +6 and +8.

Industrial Information

The noble gases do have certain important industrial functions.

Helium is used by divers to dilute the oxygen they breathe.

Argon is widely used to provide an inert atmosphere for high-temperature metallurgical processes.

Neon and argon are used for filling discharge tubes.

Further Information

For further information look up the individual elements.

Data

	Atomic Number	Relative Atomic Mass	Boiling Point/K
He	2	4.003	4.216
Ne	10	20.180	27.10
Ar	18	39.948	87.29
Kr	36	83.80	120.85
Xe	54	131.29	166.1

	1st Ionisation Energy/kJ mol^{-1}	Atomic radius/nm
He	2372.3	0.128
Ne	2080.6	0.160
Ar	1520.4	0.174
Kr	1350.7	0.189
Xe	1170.4	0.218

d-Block Elements

General Features

The d-block elements are located between Group II and Group III in the Periodic Table.

They are so called because a d-subshell is being filled. Here only the 1st row, from scandium to zinc, will be considered.

Appearance

Most of the d-block elements are considered to be metals, with a common lustrous metallic appearance.

General Reactivity

These elements have d electrons in their valence shells, and this gives them different characteristics to other metals in the Periodic Table. They each exist in several oxidation states except scandium and zinc; many of their compounds are coloured; and they readily form complexes by acting as Lewis acids.

Occurrence and Extraction

The 1st 6 elements, scandium to iron, occur mainly as the oxides in various mineral deposits. The most abundant of these is iron, found chiefly in magnetite and haematite, both commonly known as iron ore. The remainder of the elements occur mainly as sulphides such as zinc blende.

Each element is extracted from the appropriate mineral by various extraction methods. The extraction of iron, however, is of immense importance as steel - basically a mixture of iron and carbon - is used in greater quantities world-wide than any other metal. Steel is produced from iron ore in 2 main stages:

(1) a blast furnace produces impure iron from iron ore

(2) the impure iron is then purified and alloyed with other metals to produce steel.

Physical Properties

All these elements are hard, rigid and have good thermal and electrical conductivities. They have high melting and boiling points.

Chemical Properties

The chemistry of the d-block elements is governed by the fact that most exhibit several oxidation numbers. This is because the energies of all the d electrons is very similar. The d electrons also confer properties on these elements not found elsewhere:

- they easily form complexes

- their complexes are often coloured

- some complexes are paramagnetic

- they make good catalysts.

The chemical properties of these elements and their many complexes is extensive, and not suitable for further study here.

Oxidation States

As stated earlier, most of the d-block elements exist in several oxidation states - for example, the oxidation number of iron can be 0, +2, +3 and +6. The widest range of oxidation numbers is for manganese, which has a lowest oxidation number of 0 and a highest oxidation number of +7. There are general tendencies concerning the oxidation numbers:

(a) the 1st and last elements, scandium and zinc, have only 1 oxidation number.

(b) all the elements except zinc can have oxidation number +3

(c) all the elements except scandium can have oxidation number +2

(d) from scandium to manganese, the highest oxidation number = the number of 4s electrons + the number of 3d electrons

(e) from manganese to zinc, low oxidation numbers are common.

Industrial Information

The d-block elements are used in many thousands of applications.

Iron is the most widely used element because it is converted to steel, which consists of iron with 0.2 - 1.7% carbon. The addition of carbon hardens the iron and gives it better resistance to corrosion. Special steels can be prepared by the addition of small quantities of other elements - stainless steel contains 18% chromium and 8% nickel. Iron and steel are extensively used in our society.

Other important uses of some of these elements include titanium in aircraft and spaceship manufacture. Titanium is less dense than other d-block elements, and this lightness, coupled with its extra hardness, make it more suitable than aluminium in high-flying aircraft and space vessels.

Chromium is often used for electroplating, and alloyed with nickel to make nichrome - used in electrical components as its electrical resistance hardly varies with temperature.

Copper is used as protective sheeting as it is more resistant to oxidation than other elements. The green patina that forms on exposure to air is copper (II) carbonate, sulphate or chloride. Copper is also used in electrical cables.

Zinc is used in galvanising and in alloys.

Further Information

For further information look up the individual elements.

Data

Atomic Radius/nm

Sc	0.160	Y	0.177	La	0.187
Ti	0.144	Zr	0.159	Hf	0.156
V	0.131	Nb	0.142	Ta	0.143
Cr	0.124	Mo	0.136	W	0.137
Mn	0.136	Tc	0.135	Re	0.137
Fe	0.124	Ru	0.132	Os	0.133
Co	0.125	Rh	0.134	Ir	0.135
Ni	0.124	Pd	0.137	Pt	0.137
Cu	0.127	Ag	0.144	Au	0.144
Zn	0.133	Cd	0.148	Hg	0.150

Actinium Ac

General Information

Discovery

Actinium was discovered by A. Debierne in 1899 in Paris, France.

Appearance

Actinium is a soft, silvery-white metal which glows in the dark.

Source

Actinium occurs naturally in uranium minerals. It is made by the neutron bombardment of the radium isotope ^{226}Ra.

Uses

Actinium is a very powerful source of alpha rays, but is rarely used outside research.

Biological Role

Actinium has no known biological role. It is toxic due to its radioactivity.

General Information

Actinium reacts with water to evolve hydrogen gas. Its chemical properties have been little studied.

Physical Information

Atomic Number	89
Relative Atomic Mass (^{12}C=12.000)	227 (radioactive)
Melting Point/K	1320
Boiling Point/K	3470
Density/kg m^{-3}	10060 (293K)
Ground State Electron Configuration	[Rn]6d^{1}7s^2

Key Isotopes

nuclide	^{225}Ac	^{227}Ac	^{228}Ac
atomic mass		227.03	
natural abundance	0%	trace	trace
half-life	10 days	21.6 yrs	6.13 h

Ionisation Energies/kJ mol^{-1}

M - M$^+$	499
M$^+$ - M^{2+}	1170
M^{2+} - M^{3+}	1900
M^{3+} - M^{4+}	4700
M^{4+} - M^{5+}	6000
M^{5+} - M^{6+}	7300
M^{6+} - M^{7+}	9200
M^{7+} - M^{8+}	10500
M^{8+} - M^{9+}	11900
M^{9+} - M^{10+}	15800

Other Information

Enthalpy of Fusion/kJ mol^{-1} 14.2
Enthalpy of Vaporisation/kJ mol^{-1} 293

Oxidation States   Ac^{O}, Ac^{III}

Covalent Bonds /kJ mol^{-1}
not applicable

Aluminium   *Al*

General Information

Discovery

Aluminium was first prepared in an impure form by Hans Christian Oersted in Copenhagen in 1825, and isolated as an element in 1827 by Wohler.

Appearance

Aluminium is a hard and strong, silvery-white metal. An oxide film prevents it from reacting with air and water.

Source

Aluminium is not found free in nature, but is the most abundant metal in the earth's crust (8.1%) in the form of minerals such as bauxite and cryolite. Most commercially produced aluminium is obtained by the Bayer process of refining bauxite. In this process the bauxite is refined to pure aluminium oxide, which is then electrolytically reduced to pure aluminium.

Uses

Aluminium is used in an enormous variety of products, due to its particular properties. It is light, non-toxic, has a high thermal conductivity, has excellent corrosion resistance, and can be easily cast, machined and formed. It is also non-magnetic and non-sparking. It is the second most malleable metal and the sixth most ductlie. It is therefore extensively used for kitchen utensils, outside building decoration and in any area where a strong, light, easily constructed material is needed.

The electrical conductivity of aluminium is about 60% that of copper per unit area of cross-section, but it is nevertheless used in electrical transmission lines because of its light weight. Alloys of aluminium with copper, manganese, magnesium and silicon are of vital importance in the construction of aeroplanes and rockets.

Aluminium, when evaporated in a vacuum, forms a highly reflective coating for both light and heat which does not deteriorate as does a sliver coating. These aluminium coatings are used for telescope mirrors, in decorative paper, packages, toys and have many other uses.

Biological Role

Aluminium has no known biological role. It can be accumulated in the body from daily intake and has recently been implicated as a potential causative factor in Alzheimer's disease (senile dementia).

General Information (continued)

The ancient Greeks and Romans used alum (aluminium oxide) in medicine as an astringent, and in dyeing as a mordant. Sir Humphry Davy proposed the name aluminum for the element, which was undiscovered at the time, and later agreed to change it to aluminium.

Aluminium oxide, alumina, occurs naturally as ruby, sapphire, corundum and emery, and is used in glass-making and refractories.

Aluminium has largely replaced copper in cooking utensils, although it is less attractive. This is not only because it is cheaper and lighter, but also because it does not decompose vitamins during cooking. However, there is some controversy over the absorption of aluminium by the food, and potential serious health effects (see Biological Role).

Physical Information

Atomic Number	13
Relative Atomic Mass ($^{12}C=12.000$)	26.982
Melting Point/K	933.52
Boiling Point/K	2740
Density/kg m^{-3}	2698 (293K)
Ground State Electron Configuration	[Ne]$3s^2 3p^1$
Electron Affinity(M-M$^-$)/kJ mol^{-1}	44

Key Isotopes

nuclide	^{26}Al	^{27}Al
atomic mass	25.986	26.982
natural abundance	0%	100%
half-life	7.4×10^5 yrs	stable

Ionisation Energies/kJ mol^{-1}	
M - M$^+$	577.4
M$^+$ - M^{2+}	1816.6
M^{2+} - M^{3+}	2744.6
M^{3+} - M^{4+}	11575
M^{4+} - M^{5+}	14839
M^{5+} - M^{6+}	18376
M^{6+} - M^{7+}	23293
M^{7+} - M^{8+}	27457
M^{8+} - M^{9+}	31857
M^{9+} - M^{10+}	38459

Physical Information

Enthalpy of Fusion/kJ mol^{-1}	10.67
Enthalpy of Vaporisation/kJ mol^{-1}	290.8

Oxidation States

main	Al^{III}
others	Al^{O}, Al^{I}

Covalent Bonds /kJ mol^{-1}

Al-H	285
Al-C	225
Al-O	585
Al-F	665
Al-Cl	498
Al-Al	200

Americium   *Am*

General Information

Discovery

Americium was discovered by G.T. Seaborg, R.A. James, L.O. Morgan and A. Ghiorso in 1944 in Chicago, USA.

Appearance

Americium is a radioactive, silvery metal. It tarnishes slowly in dry air at room temperature.

Source

Americium can be prepared chemically by the reduction of americium (III) fluoride with barium, or americium (IV) oxide with lanthanum. However, it is produced in nuclear reactors by the neutron bombardment of plutonium, and this is the greatest source of the element.

Uses

Americium has few uses. It is of interest as it is part of the decay sequence that occurs in nuclear power production.

Biological Role

Americium has no known biological role. It is toxic due to its radioactivity.

General Information

Americium is attacked by air, steam and acids, but not by alkalis.

Physical Information

Atomic Number	95
Relative Atomic Mass ($^{12}C=12.000$)	243 (radioactive)
Melting Point/K	1267
Boiling Point/K	2880
Density/kg m^{-3}	13670 (293K)
Ground State Electron Configuration	$[Rn]5f^7 7s^2$

Key Isotopes

nuclide	^{241}Am	^{243}Am
atomic mass	241.06	243.06
natural abundance	0%	0%
half-life	458 years	7.4×10^3 yrs

Ionisation Energies/kJ mol^{-1}

M − M$^+$	578.2
M$^+$ − M^{2+}	
M^{2+} − M^{3+}	
M^{3+} − M^{4+}	
M^{4+} − M^{5+}	
M^{5+} − M^{6+}	
M^{6+} − M^{7+}	
M^{7+} − M^{8+}	
M^{8+} − M^{9+}	
M^{9+} − M^{10+}	

Other Information

Enthalpy of Fusion/kJ mol^{-1}	14.4
Enthalpy of Vaporisation/kJ mol^{-1}	238.5

Oxidation States

main oxidation state   Am^{III}

others   Am^{II}, Am^{IV}, Am^{V}, Am^{VI}

Covalent Bonds /kJ mol^{-1}

not applicable

Antimony   Sb

General Information

Discovery

Antimony was probably known to ancient civilisations and was certainly known as a metal at the beginning of the 17th century.

Appearance

Antimony exists as two allotropes, of which the metal is the usual form. This is extremely brittle, with a bright silvery colour and a hard, crystalline nature. The second allotropic form is a grey powder.

Source

Antimony is not an abundant element but is found in small quantites in over 100 mineral species. It can be found as the native metal, but more frequently as antimony (III) sulphide from which it is extracted for commercial use. This is done by roasting the antimony (III) sulphide to the oxide, and then reducing with carbon or iron.

Uses

Antimony is widely used in alloys, especially with lead in order to improve its hardness and mechanical strength, and in this form is used in batteries. Antimony is also used in semiconductor technology in making infra-red detectors and diodes. Other uses include type metal, bullets and cable sheathing.

Antimony compounds are used in manufacturing flame-proof compounds, paints, enamels, glass and pottery.

Biological Role

Antimony and many of its compounds are toxic.

General Information

Antimony exists as two allotropic forms. The normal form is metallic and stable; the other is known as the amorphous grey form.

Antimony is stable in air and is not attacked by dilute acids or alkalis. It is not acted upon by air at room temperature, but burns brilliantly when heated with the formation of white fumes of antimony (III) oxide.

Physical Information

Atomic Number	51
Relative Atomic Mass ($^{12}C=12.000$)	121.75
Melting Point/K	903.9
Boiling Point/K	1908
Density/kg m^{-3}	6691 (293K)
Ground State Electron Configuration	[Kr]4d^{10}5s^{2}5p^3
Electron Affinity(M-M$^-$)/kJ mol^{-1}	101

Key Isotopes

nuclide	^{121}Sb	^{122}Sb	^{123}Sb	^{124}Sb	^{125}Sb
atomic mass	120.9		122.93		
natural abundance	57.3%	0%	42.7%	0%	0%
half-life	stable	2.8 days	stable	60.4 days	2.71 yrs

Ionisation Energies/kJ mol^{-1}

M - M$^+$	833.7
M$^+$ - M^{2+}	1794
M^{2+} - M^{3+}	2443
M^{3+} - M^{4+}	4260
M^{4+} - M^{5+}	5400
M^{5+} - M^{6+}	10400
M^{6+} - M^{7+}	12700
M^{7+} - M^{8+}	15200
M^{8+} - M^{9+}	17800
M^{9+} - M^{10+}	20400

Other Information

Enthalpy of Fusion/kJ mol^{-1}	20.9
Enthalpy of Vaporisation/kJ mol^{-1}	165.8

Oxidation States

main	Sb^{III}, Sb^{V}
other	Sb^{-III}

Covalent Bonds /kJ mol^{-1}

Sb-H	257
Sb-C	215
Sb-O	314
Sb-F	389
Sb-Cl	313
Sb-Sb	299

Argon   *Ar*

General Information

Discovery

Argon was discovered in 1894 by Lord Rayleigh and Sir William Ramsey in the UK, although its presence in air was suspected by Cavendish in 1785.

Appearance

Argon is a colourless, odourless gas.

Source

The atmosphere contains 0.94% argon. It is obtained commercially from liquid air.

Uses

Argon is used in electric light bulbs and fluorescent tubes at a pressure of about 3 mm. Industrially, it is used as an inert gas shield for arc welding, and as a blanket for the production of titanium and other reactive elements.

Biological Role

Argon has no known biological role.

General Information

Argon is considered to be a very inert gas and does not form true compounds as do others in the same Group. However, it does form clathrates with water and quinol in which the argon atoms are trapped inside a lattice of the other molecules.

Physical Information

Atomic Number	18
Relative Atomic Mass ($^{12}C=12.000$)	39.948
Melting Point/K	83.78
Boiling Point/K	87.29
Density/kg m^{-3}	1.783 (273K)
Ground State Electron Configuration	$[Ne]3s^23p^6$
Electron Affinity(M-M$^-$)/kJ mol^{-1}	-35

Key Isotopes

nuclide	^{36}Ar	^{37}Ar	^{38}Ar	^{39}Ar	^{40}Ar
atomic mass	35.968	36.967	37.963	38.964	39.962
natural abundance	0.337%	0%	0.063%	0%	99.6%
half-life	stable	35 days	stable	269 yrs	stable

Ionisation Energies/kJ mol^{-1}

M – M$^+$	1520.4
M$^+$ – M^{2+}	2665.2
M^{2+} – M^{3+}	3928
M^{3+} – M^{4+}	5770
M^{4+} – M^{5+}	7238
M^{5+} – M^{6+}	8811
M^{6+} – M^{7+}	12021
M^{7+} – M^{8+}	13844
M^{8+} – M^{9+}	40759
M^{9+} – M^{10+}	46186

Other Information

Enthalpy of Fusion/kJ mol^{-1} 1.21
Enthalpy of Vaporisation/kJ mol^{-1} 6.53

Oxidation State   Ar^{0}

Covalent Bonds /kJ mol^{-1}

not applicable

Arsenic — As

General Information

Discovery

Arsenic was discovered in 1250 A.D. by A. Magnus, and first prepared by Schroeder in 1649.

Appearance

Arsenic is a steel grey, brittle, crystalline metalloid.

Source

The commonest arsenic-containing mineral is mispickel, and others include realgar and orpiment. Arsenic can also be found in the native state. It can be obtained from mispickel by heating, which causes the arsenic to sublime and leaves the iron (II) sulphide.

Uses

Arsenic is used in bronzing, pyrotechny and for hardening shot. It is increasingly being used as a doping agent in solid state devices.

Biological Role

Arsenic may be an essential element, but it is certainly toxic in small doses and also a suspected carcinogen. Calcium and lead arsenic compounds are used as agricultural poisons.

General Information

Arsenic has several allotropes. The most common is grey arsenic, which tarnishes and burns in oxygen. It resists attack by acids, alkalis and water but is attacked by hot acids and molten sodium hydroxide. When heated, it sublimes.

Physical Information

Atomic Number	33
Relative Atomic Mass ($^{12}C=12.000$)	74.923
Melting Point/K	1090
Boiling Point/K	889
Density/kg m^{-3}	5780 (293K)
Ground State Electron Configuration	[Ar]$3d^{10}4s^{2}4p^{3}$
Electron Affinity(M-M$^-$)/kJ mol^{-1}	77

Key Isotopes

nuclide	^{73}As	^{74}As	^{75}As	^{76}As
atomic mass	72.924	73.924	74.922	75.922
natural abundance	0%	0%	100%	0%
half-life	80.3 days	17.9 days	stable	26.5 h

Ionisation Energies/kJ mol^{-1}

M − M$^+$	947
M$^+$ − M^{2+}	1798
M^{2+} − M^{3+}	2735
M^{3+} − M^{4+}	4837
M^{4+} − M^{5+}	6042
M^{5+} − M^{6+}	12305
M^{6+} − M^{7+}	15400
M^{7+} − M^{8+}	18900
M^{8+} − M^{9+}	22600
M^{9+} − M^{10+}	26400

Other Information

Enthalpy of Fusion/kJ mol^{-1}	27.7
Enthalpy of Vaporisation/kJ mol^{-1}	31.9

Oxidation States

main	As^{III}, As^{V}
others	As^{-III}

Covalent Bonds /kJ mol^{-1}

As-H	245
As-C	200
As-O	477
As-F	464
As-Cl	293
As-As	348

Astatine *At*

General Information

Discovery

Astatine was synthesised in 1940 by D.R. Corson, K.R. MacKenzie and E. Serge in California, USA, by bombarding bismuth with alpha particles.

Source

Astatine can be obtained in various ways, but not in weighable amounts. The usual method of preparation is neutron bombardment of ^{209}Bi to produce ^{211}At.

Biological Role

Astatine has no known biological role. It is toxic due to its radioactivity.

General Information

The mass spectrometer has been used to confirm that this highly radioactive halogen behaves chemically like other halogens, particularly iodine.

Physical Information

Atomic Number	85
Relative Atomic Mass (^{12}C=12.000)	210 (radioactive)
Melting Point/K	575
Boiling Point/K	610
Ground State Electron Configuration	[Xe]$4f^{14}5d^{10}6s^{2}6p^{5}$
Electron Affinity(M-M$^-$)/kJ mol^{-1}	256

Key Isotopes

nuclide	^{210}At	^{211}At
atomic mass		210.99
natural abundance	0%	0%
half-life	8.3 h	7.21 h

Ionisation Energies/kJ mol^{-1}

M − M$^+$	930
M$^+$ − M^{2+}	1600
M^{2+} − M^{3+}	2900
M^{3+} − M^{4+}	4000
M^{4+} − M^{5+}	4900
M^{5+} − M^{6+}	7500
M^{6+} − M^{7+}	8800
M^{7+} − M^{8+}	13300
M^{8+} − M^{9+}	15400
M^{9+} − M^{10+}	17700

Other Information

Enthalpy of Fusion/kJ mol^{-1} 23.8

Oxidation States   At^{-I}, At^{I}, At^{III}

Covalent Bonds /kJ mol^{-1}

At−At   110

Barium — Ba

General Information

Discovery

Barium was discovered by Sir Humphry Davy in 1808 in London.

Appearance

Barium is a relatively soft, silvery-white metal resembling lead. It oxidises very easily and is therefore stored under petroleum or in an inert gas atmosphere.

Source

Barium occurs only in combination with other elements, chiefly in the ores barytes and witherite. It can be prepared by electrolysis of the chloride, or by heating barium oxide with aluminium.

Uses

Barium is not an extensively used element. The best-known use is in the form of barium sulphate, which can be drunk as a medical cocktail to outline the stomach and intestines for medical examination. The sulphate is also used in paint and in glassmaking.

Barium carbonate has been used as a rat poison. Barium nitrate gives fireworks a green colour.

Biological Role

Barium and all its compounds that are water or acid soluble are toxic.

General Information

Barium is attacked by air, and decomposed by water and alcohol.

Physical Information

Atomic Number	56
Relative Atomic Mass ($^{12}C=12.000$)	137.33
Melting Point/K	1002
Boiling Point/K	1910
Density/kg m^{-3}	3594 (293K)
Ground State Electron Configuration	[Xe]6s^2
Electron Affinity(M-M$^-$)/kJ mol^{-1}	-46

Key Isotopes

nuclide	^{130}Ba	^{132}Ba	^{133}Ba	^{134}Ba
atomic mass	129.9	131.9		133.9
natural abundance	0.106%	0.101%	0%	2.417%
half-life	stable	stable	7.2 yrs	stable

nuclide	^{136}Ba	^{137}Ba	^{138}Ba	^{140}Ba
atomic mass	135.9	136.9	137.9	
natural abundance	7.854%	11.32%	71.7%	0%
half-life	stable	stable	stable	12.7 days

Ionisation Energies/kJ mol^{-1}

M − M$^+$	502.8
M$^+$ − M^{2+}	965.1
M^{2+} − M^{3+}	3600
M^{3+} − M^{4+}	4700
M^{4+} − M^{5+}	6000
M^{5+} − M^{6+}	7700
M^{6+} − M^{7+}	9000
M^{7+} − M^{8+}	10200
M^{8+} − M^{9+}	13500
M^{9+} − M^{10+}	15100

Other Information

Enthalpy of Fusion/kJ mol^{-1} 7.66
Enthalpy of Vaporisation/kJ mol^{-1} 150.9

Oxidation States   Ba^{II}

Covalent Bonds /kJ mol^{-1}

not applicable

Berkelium *Bk*

General Information

Discovery

Berkelium was discovered by S.G. Thompson, A. Ghiorsoand G.T. Seaborg in 1949 in California, USA.

Appearance

Berkelium is a radioactive, silvery metal.

Source

Berkelium is made in milligram quantities only by the neutron bombardment of plutonium.

Uses

Because of its rarity, berkelium has no commercial or technological use at present.

Biological Role

Berkelium has no known biological role. It is toxic due to its radioactivity.

General Information

Berkelium is attacked by oxygen, steam and acids, but not by alkalis. Compounds with oxygen and the halides have been prepared, but only in minute quantities.

Physical Information

Atomic Number	97
Relative Atomic Mass ($^{12}C=12.000$)	247 (radioactive)
Melting Point/K	not available
Boiling Point/K	not available
Density/kg m^{-3}	14790 (293K)
Ground State Electron Configuration	[Rn]$5f^9 7s^2$

Key Isotopes

nuclide	^{247}Bk	^{249}Bk
atomic mass	247.07	
natural abundance	0%	0%
half-life	1.4×10^3 yrs	314 days

Ionisation Energies/kJ mol^{-1}

M - M$^+$	601
M$^+$ - M^{2+}	
M^{2+} - M^{3+}	
M^{3+} - M^{4+}	
M^{4+} - M^{5+}	
M^{5+} - M^{6+}	
M^{6+} - M^{7+}	
M^{7+} - M^{8+}	
M^{8+} - M^{9+}	
M^{9+} - M^{10+}	

Other Information

Enthalpy of Fusion/kJ mol^{-1}
not available

Enthalpy of Vaporisation/kJ mol^{-1}
not available

Oxidation State   Bk^{IV}

Covalent Bonds /kJ mol^{-1}

not applicable

Beryllium Be

General Information

Discovery

Discovered as the oxide by Vauquelin in beryl and in emeralds in 1798. The metal was isolated in 1828 by Wohler independently by the action of potassium on beryllium chloride.

Appearance

Beryllium is a metal, steel grey in colour.

Source

Beryllium is found in some 30 mineral species, the most important of which are bertrandite, beryl, chrysoberyl and phenacite. Aquamarine and emerald are precious forms of beryl.

Beryl and bertrandite are the most important commercial sources of the element and its compounds. The metal is usually prepared by reducing beryllium fluoride with magnesium metal.

Uses

Beryllium is used as an alloying agent in producing beryllium copper, which is used for springs, electrical contacts, spot-welding electrodes and non-sparking tools. It has found application as a structural material for high-speed aircraft, missiles, spacecraft and communication satellites, and is also extensively used in the space shuttle. Because beryllium is relatively transparent to X-rays, ultra-thin beryllium foil is finding use in X-ray lithography for the reproduction of microminiature integrated circuits.

Beryllium is also used in nuclear reactors as a reflector or moderator. The oxide has a very high melting point and is also used in nuclear work as well as having ceramic applications.

Biological Role

Beryllium and its salts are both toxic and carcinogenic.

General Information

Beryllium is one of the lightest of all metals, and has one of the highest melting points. Its modulus of elasticity is about one third greater than that of steel. It resists attack by concentrated nitric acid, has excellent thermal conductivity and is nonmagnetic. It has a high permeability to X-rays, and when bombarded by alpha particles produces neutrons. At ordinary temperatures it resists oxidation in air.

Physical Information

Atomic Number	4
Relative Atomic Mass ($^{12}C=12.000$)	9.012
Melting Point/K	1551
Boiling Point/K	3243 (under pressure)
Density/kg m^{-3}	1847.7 (293K)
Ground State Electron Configuration	[He]2s^2
Electron Affinity(M-M$^-$)/kJ mol^{-1}	-18

Key Isotopes

nuclide	^{7}Be	^{9}Be	^{10}Be
atomic mass	7.017	9.012	10.014
natural abundance	0%	100%	0%
half-life	53.37 days	stable	2.5x10^6 yrs

Ionisation Energies/kJ mol^{-1}

M - M$^+$	899.4
M$^+$ - M^{2+}	1757.1
M^{2+} - M^{3+}	14848
M^{3+} - M^{4+}	21006

Other Information

Enthalpy of Fusion/kJ mol^{-1}	9.80
Enthalpy of Vaporisation/kJ mol^{-1}	308.8

Oxidation States

main   Be^{II}

Covalent Bonds /kJ mol^{-1}

Be-H	226
Be-O	523
Be-F	615
Be-Cl	293

Bismuth   Bi

General Information

Discovery

Bismuth has been known since the fifteenth century, although it was often confused with tin and lead. Claude Geoffrey the Younger showed it to be distinct from lead in 1753.

Appearance

Bismuth is a white brittle metal with a pinkish tinge.

Source

Bismuth occurs as the native metal, and in ores such as bismuthinite and bismite. The major commercial source of bismuth is as a by-product of refining lead, copper, tin, silver and gold ores.

Uses

Bismuth is used widely in low-melting alloys with tin and cadmium, which are used in products such as fire detectors and extinguishers, electric fuses and solders. Otherwise bismuth does not find wide application.

Biological Role

Bismuth has no known biological role, and is non-toxic.

General Information

Bismuth is stable to oxygen and water, and dissolves in concentrated nitric acid. Its soluble salts are characterised by forming insoluble basic salts on the addition of water. This property has been used in forensic work.

Physical Information

Atomic Number	83
Relative Atomic Mass (^{12}C=12.000)	208.98
Melting Point/K	544.5
Boiling Point/K	1833
Density/kg m^{-3}	9747 (293K)
Ground State Electron Configuration	[Xe]4f^{14}5d^{10}6s^{2}6p^3
Electron Affinity(M-M$^-$)/kJ mol^{-1}	101

Key Isotopes

nuclide	^{206}Bi	^{207}Bi	^{209}Bi
atomic mass			208.98
natural abundance	0%	0%	100%
half-life	6.3 days	30.2 yrs	stable

Ionisation Energies/kJ mol^{-1}

M − M$^+$	703.2
M$^+$ − M^{2+}	1610
M^{2+} − M^{3+}	2466
M^{3+} − M^{4+}	4372
M^{4+} − M^{5+}	5400
M^{5+} − M^{6+}	8520
M^{6+} − M^{7+}	10300
M^{7+} − M^{8+}	12300
M^{8+} − M^{9+}	14300
M^{9+} − M^{10+}	16300

Other Information

Enthalpy of Fusion/kJ mol^{-1}	10.48
Enthalpy of Vaporisation/kJ mol^{-1}	179.1

Oxidation States

main	Bi^{III}
others	Bi^{-III}, Bi^{I}, Bi^{V}

Covalent Bonds /kJ mol^{-1}

Bi-H	194
Bi-C	143
Bi-O	339
Bi-F	314
Bi-Cl	285
Bi-Bi	200

Boron B

General Information

Discovery

Boron was discovered in 1808 by L.J. Lussac and L.J. Thenard in Paris, and Sir Humphry Davy in London.

Appearance

The element is a grey powder, but is not found free in nature.

Source

Boron occurs as orthoboric acid in certain volcanic spring waters, and as borates in the minerals borax and colemanite. However, by far the most important source of boron is rasorite, which is found in the Mojave desert in California. Extensive borax deposits are also found in Turkey.

High purity boron is prepared by the vapour phase reduction of boron trichloride or tribromide with hydrogen on electrically heated filaments. The impure, or amorphous, boron can be prepared by heating the trioxide with magnesium powder.

Uses

Amorphous boron is used in pyrotechnic flares to provide a distinctive green colour, and in rockets as an igniter. The most important compounds of boron are boric, or boracic acid, widely used as a mild antiseptic, and borax which serves as a cleansing flux in welding and as a water softener in washing powders. Boron compounds are also extensively used in the manufacture of borosilicate glasses. Other boron compounds show promise in treating arthritis. The isotope boron 10 is used as a control for nuclear reactors, as a shield for nuclear radiation, and in instruments used for detecting neutrons. Demand is increasing for boron filaments, a high-strength, lightweight material chiefly employed for advanced aerospace structures.

Biological Role

Elemental boron is not considered a poison, and indeed is essential to plants, but assimilation of its compounds has a cumulative toxic effect.

General Information

Elemental boron has an energy band gap of 1.50 to 1.56 eV, which is higher than that of either silicon or germanium. It has interesting optical characteristics, transmitting portions of the infra-red only. It is a poor conductor of electricity at room temperature, but a good conductor at high temperatures.

Physical Information

Atomic Number	5
Relative Atomic Mass ($^{12}C=12.000$)	10.81
Melting Point/K	2573
Boiling Point/K	3931
Density/kg m^{-3}	2340 (293K)
Ground State Electron Configuration	[He]$2s^2 2p^1$
Electron Affinity(M-M$^-$)/kJ mol^{-1}	15

Key Isotopes

nuclide	^{10}B	^{11}B
atomic mass	10.013	11.009
natural abundance	20.0%	80.0%
half-life	stable	stable

Ionisation Energies/kJ mol^{-1}

M - M$^+$	800.6
M$^+$ - M^{2+}	2427
M^{2+} - M^{3+}	3660
M^{3+} - M^{4+}	25025
M^{4+} - M^{5+}	32822

Other Information

Enthalpy of Fusion/kJ mol^{-1}	22.2
Enthalpy of Vaporisation/kJ mol^{-1}	504.5

Oxidation State B^{III}

Covalent Bonds /kJ mol^{-1}

B-H	381
B-H-B	439
B-C	372
B-O	523
B-F	644
B-Cl	444
B-B	335

Key Isotopes

nuclide	^{77}Br	^{79}Br	^{81}Br	^{82}Br
atomic mass		78.918	80.916	81.917
natural abundance	0%	50.69%	49.31%	0%
half-life	57 h	stable	stable	35.5 h

Ionisation Energies/kJ mol^{-1}

M – M$^+$	1139.9
M$^+$ – M^{2+}	2104
M^{2+} – M^{3+}	3500
M^{3+} – M^{4+}	4560
M^{4+} – M^{5+}	5760
M^{5+} – M^{6+}	8550
M^{6+} – M^{7+}	9940
M^{7+} – M^{8+}	18600
M^{8+} – M^{9+}	23900
M^{9+} – M^{10+}	28100

Other Information

Enthalpy of Fusion/kJ mol^{-1}	10.8
Enthalpy of Vaporisation/kJ mol^{-1}	30.5

Oxidation States

main	Br^{-I}, Br^{V}
others	Br^{I}, Br^{III}, Br^{IV}, Br^{VII}

Covalent Bonds /kJ mol^{-1}

Br-H	366
Br-C	285
Br-O	234
Br-F	285
Br-Br	193
Br-B	410
Br-Si	310
Br-P	264

Bromine Br

General Information

Discovery
Bromine was discovered by A.J. Balard in 1826 in Montpelier, France.

Appearance
Bromine is a red, dense liquid with a sharp, distinctive smell.

Source
Bromine is extracted from natural brine deposits in the USA and elsewhere. It was the first compound to be extracted from seawater but this is no longer economically viable as seawater contains only 65 parts per million of bromine.

Uses
Bromine is used in many areas such as agricultural chemicals, dyestuffs, chemical intermediates and flame retardants. Most is used to prepare 1,2-di-bromoethane which is used as an anti-knock agent in combustion engines.

Biological Role
Bromine has no known biological role. It has an irritating effect on the eyes and throat, and produces painful sores when in contact with the skin.

General Information
Bromine combines readily with many elements. Like chlorine, it has a natural bleaching action.

Physical Information

Atomic Number	35
Relative Atomic Mass ($^{12}C=12.000$)	79.904
Melting Point/K	265.9
Boiling Point/K	331.9
Density/kg m^{-3}	3122 (293K)
Ground State Electron Configuration	$[Ar]3d^{10}4s^24p^5$
Electron Affinity(M-M$^-$)/kJ mol^{-1}	324

Cadmium   *Cd*

General Information

Discovery

Cadmium was discovered by F. Stromeyer in 1817 in Gottingen, Germany, from an impurity in zinc carbonate.

Appearance

Cadmium is a soft, bluish-white metal which is easily cut with a knife.

Source

The only mineral containing significant quantities of cadmium is greenockite, although some is present in sphalerite. Almost all commercially produced cadmium is obtained as a by-product in the treatment of zinc, copper and lead ores.

Uses

Cadmium is used extensively in electroplating, which accounts for about 60% of its use. It is also used in many types of solder, for standard e.m.f. cells, for nickel-cadmium batteries and as a barrier to control atomic fission. It is a component of some of the lowest melting alloys, alloys with low coefficients of friction and alloys with great resistance to fatigue. Cadmium compounds are used in blue and green phosphors in colour television sets. Cadmium forms a number of compounds, the sulphide being used as an artist's pigment as it is bright yellow.

Biological Role

Cadmium is toxic, carcinogenic and teratogenic. In the past, failure to recognise the toxicity of this element caused workers to be exposed to danger in the form of solder fumes and cadmium plating baths.

General Information

Cadmium tarnishes in air, is soluble in acids but not in alkalis.

Physical Information

Atomic Number	48
Relative Atomic Mass ($^{12}C=12.000$)	112.41
Melting Point/K	594.1
Boiling Point/K	1038
Density/kg m^{-3}	8650 (293K)
Ground State Electron Configuration	[Kr]4d^{10}5s^2
Electron Affinity(M-M$^-$)/kJ mol^{-1}	-26

Key Isotopes

nuclide	^{106}Cd	^{108}Cd	^{109}Cd	^{110}Cd	^{111}Cd
atomic mass	105.91	107.9		109.9	110.9
natural abundance	1.25%	0.89%	0%	12.51%	12.81%
half-life	stable	stable	450 days	stable	stable

nuclide	^{112}Cd	^{113}Cd	^{114}Cd	^{115}Cd	^{116}Cd
atomic mass	111.9	112.9	113.9		115.9
natural abundance	0%	24.13%	12.22%	28.72%	7.47%
half-life	53.5 h	stable	stable	stable	stable

Ionisation Energies/kJ mol^{-1}

M - M$^+$	867.6
M$^+$ - M^{2+}	1631
M^{2+} - M^{3+}	3616
M^{3+} - M^{4+}	5300
M^{4+} - M^{5+}	7000
M^{5+} - M^{6+}	9100
M^{6+} - M^{7+}	11100
M^{7+} - M^{8+}	14100
M^{8+} - M^{9+}	16400
M^{9+} - M^{10+}	18800

Other Information

Enthalpy of Fusion/kJ mol^{-1}	6.11
Enthalpy of Vaporisation/kJ mol^{-1}	100

Oxidation States

main	Cd^{II}
other	Cd^{I}

Covalent Bonds /kJ mol^{-1}

not applicable

Caesium Cs

General Information

Discovery

Caesium was discovered by R. Bunsen and G.R. Kirchoff in 1860 in Heidelberg, Germany.

Appearance

Caesium is silvery-white, soft and ductile. It is liquid at room temperature.

Source

Caesium is found in the minerals pollucite and lepidolite. Pollucite is found in great quantities at Bernic Lake, Manitoba, Canada and the USA, and from this source the element can be prepared. However, most commercial production is as a by-product of lithium production.

Uses

Caesium is little used. It has a great affinity for oxygen and so is used in electron tubes, and it is also used in photoelectric cells and as a catalyst. A more interesting application is the use in atomic clocks which are accurate to 5 seconds in 300 years.

Biological Role

Caesium has no known biological role. It is non-toxic.

General Information

Caesium reacts rapidly with oxygen and explosively with water. It also reacts with ice at temperatures above 116K. The metal is characterised by a spectrum containing two bright lines in the blue along with several others in the red, yellow and green. Caesium hydroxide is the strongest base known, and can attack glass.

Physical Information

Atomic Number	55
Relative Atomic Mass ($^{12}C=12.000$)	132.91
Melting Point/K	301.6
Boiling Point/K	951.6
Density/kg m^{-3}	1873 (293K)
Ground State Electron Configuration	[Xe]6s^1
Electron Affinity(M-M$^-$)/kJ mol^{-1}	45.5

Key Isotopes

nuclide	^{133}Cs	^{134}Cs	^{135}Cs	^{137}Cs
atomic mass	132.9			
natural abundance	100%	0%	0%	0%
half-life	stable	2.05 yrs	3x10^6 yrs	30.23 yrs

Ionisation Energies/kJ mol^{-1}

M - M$^+$	375.7
M$^+$ - M^{2+}	2420
M^{2+} - M^{3+}	3400
M^{3+} - M^{4+}	4400
M^{4+} - M^{5+}	6000
M^{5+} - M^{6+}	7100
M^{6+} - M^{7+}	8300
M^{7+} - M^{8+}	11300
M^{8+} - M^{9+}	12700
M^{9+} - M^{10+}	23700

Other Information

Enthalpy of Fusion/kJ mol^{-1} 2.09
Enthalpy of Vaporisation/kJ mol^{-1} 66.5

Oxidation States   Cs^{-I}, Cs^{I}

Covalent Bonds /kJ mol^{-1}

not applicable

Calcium Ca

General Information

Discovery

Calcium was first isolated by Sir Humphry Davy in 1808 in London, although lime, or calcium oxide, was prepared by the Romans in the first century.

Appearance

Calcium is a silvery-white, relatively soft metal.

Source

Calcium is the fifth most abundant metal in the earth's crust, greater than 3% by mass. It is not found free in nature, but occurs abundantly as limestone (calcium carbonate), gypsum (calcium sulphate), fluorite (calcium fluoride) and apatite (calcium chloro- or fluoro-phosphate). Calcium is prepared commercially by the electrolysis of fused calcium chloride to which calcium fluoride is added to lower the melting point.

Uses

Calcium and its compounds are widely used. Quicklime (calcium oxide), which is made by heating limestone and can be changed into slaked lime by the addition of water, is a substance often used by the chemical industry. It has the advantage of being cheap and readily available. When mixed with sand it takes up carbon dioxide from the air and hardens as mortar and plaster. Calcium from limestone is an important constituent of Portland Cement. Calcium is also used as a reducing agent in preparing other metals such as thorium and uranium, and as an alloying agent for aluminium, beryllium, copper, lead and magnesium alloys.

Biological Role

Calcium is an essential constituent of cells, teeth and bones. The normal amount found in an adult is over four kilograms, located mostly in the teeth and bones.

General Information

Calcium readily forms a white coating of nitride in air, it reacts with water and burns with a yellow-red flame, forming mostly the nitride. Calcium carbonate is soluble in water containing carbon dioxide, and this causes hardness in water. This calcium carbonate is also the constituent of stalactites and stalagmites in caves where water drips slowly and evaporates in situ.

Physical Information

Atomic Number	20
Relative Atomic Mass ($^{12}C=12.000$)	40.078
Melting Point/K	1112
Boiling Point/K	1757
Density/kg m^{-3}	1550 (293K)
Ground State Electron Configuration	[Ar]4s^2
Electron Affinity(M-M$^-$)/kJ mol^{-1}	−186

Key Isotopes

nuclide	^{40}Ca	^{42}Ca	^{43}Ca	^{44}Ca	^{45}Ca
atomic mass	39.963	41.959	42.959	43.955	44.956
natural abundance	96.94%	0.647%	0.135%	2.086%	0%
half-life	stable	stable	stable	stable	165 days

nuclide	^{46}Ca	^{47}Ca	^{48}Ca
atomic mass	45.954	46.954	47.952
natural abundance	0.004%	0%	0.187%
half-life	stable	4.53 hours	stable

Ionisation Energies/kJ mol^{-1}

M − M$^+$	589.7
M$^+$ − M^{2+}	1145
M^{2+} − M^{3+}	4910
M^{3+} − M^{4+}	6474
M^{4+} − M^{5+}	8144
M^{5+} − M^{6+}	10496
M^{6+} − M^{7+}	12320
M^{7+} − M^{8+}	14207
M^{8+} − M^{9+}	18191
M^{9+} − M^{10+}	20385

Other Information

Enthalpy of Fusion/kJ mol^{-1}	9.33
Enthalpy of Vaporisation/kJ mol^{-1}	150.6

Oxidation State Ca^{II}

Covalent Bonds /kJ mol^{-1}

not applicable

Californium　　　*Cf*

General Information

Discovery

Californium was discovered by S.G. Thompson, K. Street, A. Ghiorso and G.T. Seaborg in 1950 in California, USA.

Appearance

Californium is a radioactive, silvery metal.

Source

Californium did not exist in weighable amounts until ten years after its discovery. The usual method of preparation, producing milligram amounts only, is by neutron bombardment of plutonium.

Uses

Californium is a very strong neutron emitter. It is therefore used as a portable neutron source for the discovery of metals such as gold and silver. One isotope, ^{252}Cf, is used in cancer therapy.

Biological Role

Californium has no known biological role. It is toxic due to its radioactivity.

General Information

Californium is attacked by oxygen, steam and acids, but not by alkalis.

Physical Information

Atomic Number	98
Relative Atomic Mass (^{12}C=12.000)	251 (radioactive)
Melting Point/K	not available
Boiling Point/K	not available
Density/kg m^{-3}	not available
Ground State Electron Configuration	[Rn]5f^{10}7s^2
Electron Affinity(M-M$^-$)/kJ mol^{-1}	not available

Key Isotopes

nuclide	^{249}Cf	^{251}Cf	^{252}Cf
atomic mass	249.07		
natural abundance	0%	0%	0%
half-life	360 yrs	900 yrs	2.65 yrs

Ionisation Energies/kJ mol^{-1}

M − M$^+$	608
M$^+$ − M^{2+}	
M^{2+} − M^{3+}	
M^{3+} − M^{4+}	
M^{4+} − M^{5+}	
M^{5+} − M^{6+}	
M^{6+} − M^{7+}	
M^{7+} − M^{8+}	
M^{8+} − M^{9+}	
M^{9+} − M^{10+}	

Other Information

Enthalpy of Fusion/kJ mol^{-1}

not available

Enthalpy of Vaporisation/kJ mol^{-1}

not available

Oxidation States

main	Cf^{III}
others	Cf^{II}, Cf^{IV}

Covalent Bonds /kJ mol^{-1}

not applicable

Carbon   C

General Information

Discovery

Carbon is an element of prehistoric discovery and is widely distributed in nature.

Appearance

Carbon can exist as either black graphite (known as charcoal in the powdered form) or as the colourless gem diamond.

Source

Carbon is found in abundance in the sun, stars, comets and atmospheres of most planets.

Graphite is found naturally in many locations. Diamond is found in the form of microscopic crystals in some meteorites. Natural diamonds are found in the mineral kimberlite, sources of which are in South Africa, Arkansas and elsewhere. Diamonds are now also being recovered from the ocean floor off the Cape of Good Hope.

About 30% of all industrial diamonds used in the United States are made synthetically.

Carbon is found in combination in hydrocarbons (methane gas, oil and coal), and carbonates (limestone and dolomite).

Uses

Carbon is unique among the elements in the vast number and variey of compounds it can form. With hydrogen, oxygen, nitrogen and other elements it forms very large numbers of compounds, carbon atom often being linked to carbon atom.

This ability to form chains is unique to carbon, and is thought to be an important reason for the dependance of life on this element. It is also an indispensable source of such varied everyday products as nylon and petrol, perfume and plastics, shoe polish, DDT and TNT.

Biological Role

Carbon is the basis of all life as part of the DNA molecule. There are more than a million known carbon compounds, many thousands of which are vital to organic and life processes.

General Information (continued)

Carbon is found free in nature in three allotropic forms; amorphous, graphite and diamond. A fourth form known as "white" carbon, is now thought to exist. Graphite is one of the softest known materials and diamond one of the hardest. This difference is purely because of the arrangement of atoms in each of the two forms. In graphite, hexagonal rings are joined together to form sheets, and the sheets lie one on top of the other. In diamond, the atoms are arranged tetrahedrally in a vast continuous array. "White" carbon is a transparent birefringent material produced during the sublimation of graphite at low pressures.

In 1961 the International Union of Pure and Applied Chemistry adopted the isotope carbon-12 as the basis for atomic masses.

Carbon-14, an isotope with a half-life of 5730 years, has been widely used to date materials such as wood, archeological specimens etc.

Physical Information

Atomic Number	6
Relative Atomic Mass ($^{12}C=12.000$)	12.011
Melting Point/K	3820 (diamond)
Boiling Point/K	5100 (sublimes)
Density/kg m^{-3}	3513 (diam.) 2260
Ground State Electron Configuration	[He]$2s^22p^2$
Electron Affinity(M-M$^-$)/kJ mol^{-1}	121

Key Isotopes

nuclide	^{12}C	^{13}C	^{14}C
atomic mass	12.000	13.003	14.003
natural abundance	98.90%	1.10%	trace
half-life	stable	stable	5730 yrs

Ionisation Energies/kJ mol^{-1}	
M - M$^+$	1086.2
M$^+$ - M^{2+}	2352
M^{2+} - M^{3+}	4620
M^{3+} - M^{4+}	6222
M^{4+} - M^{5+}	37827
M^{5+} - M^{6+}	47270

Other Information

Enthalpy of Fusion/kJ mol^{-1}	105.0
Enthalpy of Vaporisation/kJ mol^{-1}	710.9

Oxidation States

This concept is rarely used in discussing carbon in its compounds because of subtleties of bonding. However, in single compounds it can be regarded as having oxidation states of C^{-IV}, C^{II}, C^{IV}

Covalent Bonds /kJ mol^{-1}

C-H	411
C-C	348
C=C	614
C≡C	839
C=N	615
C≡N	891
C=O	745
C≡O	1074

Cerium   *Ce*

General Information

Discovery

Cerium was discovered by J.J. Berzelius and W. Hisinger in 1803 in Vestmanland, Sweden. It was first isolated by Hillebrand and Norton in 1875, in Washington, USA.

Appearance

Cerium is an iron-grey, lustrous, malleable metal. It oxidises easily at room temperature.

Source

Cerium is the most abundant of the lanthanides and is found in a number of minerals, chiefly bastnaesite (found in Southern California) and monazite (found in India and Brazil). Metallic cerium can be prepared by two methods. The first is the metallothermic reduction of cerium (III) fluoride with calcium, used to produce high-purity cerium. The second is the electrolysis of molten cerium (III) chloride.

Uses

Cerium is the major component of misch-metal alloy (just under 50%), which is used extensively in the manufacture of pyrophoric alloys for products such as cigarette lighters. Cerium (III) oxide is used as a catalyst in self-cleaning ovens, incorporated into oven walls to prevent the build-up of cooking residues. It is also a promising new petroleum-cracking catalyst.

Biological Role

Cerium has no known biological role.

General Information

Cerium tarnishes in air and reacts rapidly with water, especially when hot. It burns when heated. It is attacked by alkali solutions and all acids. The pure metal is likely to ignite when scratched with a knife.

Cerium is interesting because of its variable electronic structure. The energy of the inner 4f level is nearly the same as that of the 6s level, and this gives rise to variable occupancy of these two levels and subsequent variable oxidation states.

Physical Information

Atomic Number	58
Relative Atomic Mass ($^{12}C=12.000$)	140.12
Melting Point/K	1072
Boiling Point/K	3699
Density/kg m^{-3}	6773 (298K)
Ground State Electron Configuration	[Xe]4f^{2}6s^2
Electron Affinity(M-M$^-$)/kJ mol^{-1}	50

Key Isotopes

nuclide	^{136}Ce	^{138}Ce	^{139}Ce	^{140}Ce
atomic mass		137.9		139.9
natural abundance	0.19%	0.25%	0%	88.48%
half-life	stable	stable	140 days	stable

nuclide	^{141}Ce	^{142}Ce	^{143}Ce	^{144}Ce
atomic mass		141.9		
natural abundance	0%	11.08%	0%	0%
half-life	32.5 days	stable	33 h	284.9 h

Ionisation Energies/kJ mol^{-1}

M - M$^+$	527.4
M$^+$ - M^{2+}	1047
M^{2+} - M^{3+}	1949
M^{3+} - M^{4+}	3547
M^{4+} - M^{5+}	6800
M^{5+} - M^{6+}	8200
M^{6+} - M^{7+}	9700
M^{7+} - M^{8+}	11800
M^{8+} - M^{9+}	13200
M^{9+} - M^{10+}	14700

Other Information

Enthalpy of Fusion/kJ mol^{-1}	8.87
Enthalpy of Vaporisation/kJ mol^{-1}	398

Oxidation States

main	Ce^{III}
others	Ce^{IV}

Covalent Bonds /kJ mol^{-1}

not applicable

Chlorine Cl

General Information

Discovery

Chlorine was discovered in 1774 by C.W. Scheele in Uppsala, Sweden. He thought it contained oxygen and it was Davy who recognised it as an element and named it chlorine in 1810.

Appearance

Chlorine is a greenish-yellow, dense gas with a sharp smell.

Source

Chlorine is not found free in nature but combined chiefly with sodium as sodium chloride in common salt and the minerals carnallite and sylvite.

Chlorine is produced commercially by the electrolysis of sodium chloride.

Uses

Chlorine is widely used in many different areas. It is used in the production of safe drinking water and many consumer products such as paper, dyestuffs, textiles, petroleum products, medicines, antiseptics, insecticides, foodstuffs, solvents, paints and plastics. It is also used to produce chlorates, chloroform, carbon tetrachloride and bromine. A further substantial use for this element is in organic chemistry, both as an oxidising agent and in substitution reactions.

Biological Role

The chloride ion is essential to life. Chlorine gas is a respiratory irritant, which can be fatal after a few deep breaths. It was used as a war gas in 1915. Chlorine liquid burns the skin.

Physical Information

Atomic Number	17
Relative Atomic Mass ($^{12}C=12.000$)	35.453
Melting Point/K	172.17
Boiling Point/K	239.18
Density/kg m^{-3}	3.214 (273K)
Ground State Electron Configuration	[Ne] $3s^2 3p^5$
Electron Affinity(M-M$^-$)/kJ mol^{-1}	348

Key Isotopes

nuclide	^{35}Cl	^{36}Cl	^{37}Cl
atomic mass	34.969	35.980	36.966
natural abundance	75.77%	0%	24.23%
half-life	stable	3.1×10^5 yrs	stable

Ionisation Energies/kJ mol^{-1}

M - M$^+$	1251.1
M$^+$ - M^{2+}	2297
M^{2+} - M^{3+}	3826
M^{3+} - M^{4+}	5158
M^{4+} - M^{5+}	6540
M^{5+} - M^{6+}	9362
M^{6+} - M^{7+}	11020
M^{7+} - M^{8+}	33610
M^{8+} - M^{9+}	38600
M^{9+} - M^{10+}	43960

Other Information

Enthalpy of Fusion/kJ mol^{-1} 6.41
Enthalpy of Vaporisation/kJ mol^{-1} 20.403

Oxidation States

main   Cl^{-I}, Cl^{VII}
others   Cl^{I}, Cl^{III}, Cl^{IV}, Cl^{V}, Cl^{VI}

Covalent Bonds /kJ mol^{-1}

Cl-O	206
Cl-Cl	242
Cl-F	257

Chromium   *Cr*

General Information

Discovery

Chromium was discovered in 1780 by N.L. Vanquelin in Paris.

Appearance

Chromium is a blue-white, hard metal, capable of taking a high polish.

Source

Chromium is found principally in the ore chromite, which is found in many places including the former USSR, Turkey, Iran, Finland and the Phillipines. Chromium metal is usually produced commercially by reduction of chromium (III) oxide by aluminium, or electrolysis of chrome alum.

Uses

Chromium is used to harden steel, to manufacture stainless steel and to produce several alloys. It is also used in plating as it prevents corrosion and gives a high-lustre finish. It is also used as a catalyst.

Chromium compounds are valued as pigments for their vivid green, yellow, red and orange colours. The ruby takes its colour from chromium, and chromium added to glass imparts an emerald green colour.

Biological Role

Chromium is an essential trace element, but is carcinogenic in excess. Chromium compounds are toxic.

Physical Information

Atomic Number	24
Relative Atomic Mass ($^{12}C=12.000$)	51.996
Melting Point/K	2130
Boiling Point/K	2945
Density/kg m^{-3}	7190 (293K)
Ground State Electron Configuration	$[Ar]3d^5 4s^1$
Electron Affinity(M-M$^-$)/kJ mol^{-1}	94

Key Isotopes

nuclide	^{50}Cr	^{51}Cr	^{52}Cr	^{53}Cr	^{54}Cr
atomic mass	49.946	50.945	51.941	52.941	53.939
natural abundance	4.359%	0%	83.79%	9.50%	2.36%
half-life	stable	27.8 days	stable	stable	stable

Ionisation Energies/kJ mol^{-1}

M − M$^+$	652.7
M$^+$ − M^{2+}	1592
M^{2+} − M^{3+}	2987
M^{3+} − M^{4+}	4740
M^{4+} − M^{5+}	6690
M^{5+} − M^{6+}	8738
M^{6+} − M^{7+}	15550
M^{7+} − M^{8+}	17830
M^{8+} − M^{9+}	20220
M^{9+} − M^{10+}	23580

Other Information

Enthalpy of Fusion/kJ mol^{-1} 15.3
Enthalpy of Vaporisation/kJ mol^{-1} 341.8

Oxidation States

main   Cr^{III}

others   Cr^{-II}, Cr^{-I}, Cr^{O}, Cr^{I}, Cr^{II}, Cr^{IV}, Cr^{V}, Cr^{VI}

Covalent Bonds /kJ mol^{-1}

not applicable

Cobalt *Co*

General Information

Discovery
Cobalt was discovered by G. Brandt in 1735 in Stockholm, Sweden.

Appearance
Cobalt is a lustrous, silvery-blue, hard metal.

Source
Cobalt is found in the minerals cobaltite, smaltite and erythrite. Important ore deposits are found in Zaire, Morocco and Canada. It is also thought that the floor of the north central Pacific Ocean may have cobalt-rich deposits.

Uses
Cobalt metal is used in electroplating because of its attractive appearance, hardness and resistance to oxidation. It is alloyed with iron, nickel and other metals, and used in jet turbines and gas turbine generators.

Cobalt salts have been used for centuries to produce brilliant blue colours in porcelain, glass, pottery and enamels.

Radioactive cobalt-60 is used in the treatment of cancer.

Biological Role
Cobalt is an essential trace element, and forms part of the active site of vitamin B_{12}. Cobalt salts in small doses have been found to be effective in correcting mineral deficiencies in certain animals. Cobalt in large doses is carcinogenic. Radioactive artificial cobalt-60 is an important gamma-ray source, and is used extensively as a tracer and radiotherapeutic agent.

Physical Information

Atomic Number	27
Relative Atomic Mass (^{12}C=12.000)	58.933
Melting Point/K	1768
Boiling Point/K	3143
Density/kg m^{-3}	8900 (293K)
Ground State Electron Configuration	$[Ar]3d^7 4s^2$
Electron Affinity(M-M$^-$)/kJ mol^{-1}	102

Key Isotopes

nuclide	^{56}Co	^{57}Co	^{58}Co	^{59}Co	^{60}Co
atomic mass	55.940	56.936	57.936	58.933	59.934
natural abundance	0%	0%	0%	100%	0%
half-life	77 days	270 days	71.3 days	stable	5.26 yrs

Ionisation Energies/kJ mol^{-1}

M – M$^+$	760
M$^+$ – M^{2+}	1646
M^{2+} – M^{3+}	3232
M^{3+} – M^{4+}	4950
M^{4+} – M^{5+}	7670
M^{5+} – M^{6+}	9840
M^{6+} – M^{7+}	12400
M^{7+} – M^{8+}	15100
M^{8+} – M^{9+}	17900
M^{9+} – M^{10+}	26600

Other Information

Enthalpy of Fusion/kJ mol^{-1} 15.2
Enthalpy of Vaporisation/kJ mol^{-1} 382.4

Oxidation States

main   Co^{II}
others   Co^{-I}, Co^{O}, Co^{I}, Co^{III}, Co^{IV}, Co^{V}

Covalent Bonds /kJ mol^{-1}

not applicable

Copper   *Cu*

General Information

Discovery

Copper was known to ancient civilisations, and is said to have been mined for more than 5000 years.

Appearance

Copper is a reddish colour and takes on a bright sheen. It is malleable and ductlie.

Source

Copper metal does occur naturally, but by far the greatest source is in minerals such as chalcopyrite and bornite. Copper ores (copper sulphides, oxides and carbonates) are found in the USA and Canada, as well as several other places. From these ores and minerals copper is obtained by smelting, leaching and electrolysis.

Uses

The greatest percentage of copper used is in electrical equipment such as wiring and motors. Brass and bronze are both copper alloys and are extensively used. All American coins are now copper alloys, and gun metals also contain copper.

Copper sulphate is used widely as an agricultural poison and as an algicide in water purification. Copper compounds such as Fehling's solution are used in chemical tests for sugar detection.

Biological Role

Copper is an essential element although excess copper is toxic.

General Information

Copper is a good conductor of heat and electricity - hence its use in the electrical industry.

It is resistant to air and water but slowly weathers to the green patina of the carbonate often seen on roofs and statues.

Physical Information

Atomic Number	29
Relative Atomic Mass ($^{12}C=12.000$)	63.546
Melting Point/K	1356.6
Boiling Point/K	2840
Density/kg m^{-3}	8960 (293K)
Ground State Electron Configuration	[Ar]$3d^{10}4s^1$
Electron Affinity(M-M$^-$)/kJ mol^{-1}	118.3

Key Isotopes

nuclide	^{63}Cu	^{64}Cu	^{65}Cu	^{67}Cu
atomic mass	62.930	63.930	64.928	
natural abundance	69.17%	0%	30.83%	
half-life	stable	12.9 h	stable	61.88 h

Ionisation Energies/kJ mol^{-1}

M − M$^+$	745.4
M$^+$ − M^{2+}	1958
M^{2+} − M^{3+}	3554
M^{3+} − M^{4+}	5326
M^{4+} − M^{5+}	7709
M^{5+} − M^{6+}	9940
M^{6+} − M^{7+}	13400
M^{7+} − M^{8+}	16000
M^{8+} − M^{9+}	19200
M^{9+} − M^{10+}	22400

Other Information

Enthalpy of Fusion/kJ mol^{-1} 13.0
Enthalpy of Vaporisation/kJ mol^{-1} 306.7

Oxidation States

main   Cu^{II}
others   Cu^{-I}, Cu^{O}, Cu^{I}, Cu^{III}, Cu^{IV}

Covalent Bonds /kJ mol^{-1}

not applicable

Curium   *Cm*

General Information

Discovery

Curium was discovered by G.T. Seaborg, R.A. James and A. Ghiorso in 1944 in California, USA.

Appearance

Curium is a radioactive, silvery metal.

Source

Curium can be made in very small amounts by the neutron bombardment of plutonium. Minute amounts may exist in natural deposits of uranium.

Uses

Curium has little use outside research as it is only available in extremely small quantities.

Biological Role

Curium has no known biological role. It is toxic due to its radioactivity.

General Information

Curium is attacked by oxygen, steam and acids, but not by alkalis. Several oxides and halides of this element have been prepared.

Physical Information

Atomic Number	96
Relative Atomic Mass ($^{12}C=12.000$)	247 (radioactive)
Melting Point/K	1610
Boiling Point/K	not available
Density/kg m^{-3}	13300 (293K)
Ground State Electron Configuration	$[Rn]5f^76d^17s^2$

Key Isotopes

nuclide	^{242}Cm	^{244}Cm	^{247}Cm	^{248}Cm
atomic mass	242.06	244.06	247.07	
natural abundance	0%	0%	0%	0%
half-life	163 days	17.6yrs	1.6×10^7 yrs	4.7×10^5 yrs

Ionisation Energies/kJ mol^{-1}

M – M$^+$	581
M$^+$ – M^{2+}	
M^{2+} – M^{3+}	
M^{3+} – M^{4+}	
M^{4+} – M^{5+}	
M^{5+} – M^{6+}	
M^{6+} – M^{7+}	
M^{7+} – M^{8+}	
M^{8+} – M^{9+}	
M^{9+} – M^{10+}	

Other Information

Enthalpy of Fusion/kJ mol^{-1}

not available

Enthalpy of Vaporisation/kJ mol^{-1}

not available

Oxidation States

main   Cm^{III}
others   Cm^{II}, Cm^{IV}

Covalent Bonds /kJ mol^{-1}

not applicable

Dysprosium　　　Dy

General Information

Discovery

Dysprosium was discovered by P.E. Lecoq de Boisbaudran in 1886 in Paris, France.

Appearance

Dysprosium is a bright, hard metal with a silvery lustre.

Source

In common with many other rare earth elements, dysprosium is found in the minerals monazite and bastnaesite, and in smaller quantities in several other minerals such as xenotime and fergusonite. It can be extracted from these minerals by ion exchange and solvent extraction. It can also be prepared by the reduction of the trifluoride with calcium metal.

Uses

Dysprosium has not yet found many applications. However, it has a high thermal neutron absorption cross-section and a high melting point, and so it may be useful in nuclear control alloys. A dysprosium oxide-nickel cement is used in cooling nuclear reactor control rods, and has the property of absorbing neutrons readily without swelling or contracting under prolonged neutron bombardment.

Biological Role

Dysprosium has no known biological role, and has low toxicity.

General Information

Dysprosium is relatively stable in air at room temperature, and is readily attacked and dissolved by acids. It is soft enough to be cut with a knife.

Physical Information

Atomic Number	66
Relative Atomic Mass ($^{12}C=12.000$)	162.50
Melting Point/K	1685
Boiling Point/K	2835
Density/kg m^{-3}	8550 (293K)
Ground State Electron Configuration	$[Xe]4f^{10}6s^2$
Electron Affinity(M-M$^-$)/kJ mol^{-1}	not available

Key Isotopes

nuclide	^{156}Dy	^{158}Dy	^{160}Dy	^{161}Dy
atomic mass	155.9	157.9	159.9	160.9
natural abundance	0.06%	0.10%	2.34%	18.9%
half-life	stable	stable	stable	stable

nuclide	^{162}Dy	^{163}Dy	^{164}Dy
atomic mass	161.9	162.9	163.9
natural abundance	25.5%	24.9%	28.2%
half-life	stable	stable	stable

Ionisation Energies/kJ mol^{-1}

M - M$^+$	571.9
M$^+$ - M^{2+}	1126
M^{2+} - M^{3+}	2200
M^{3+} - M^{4+}	4001
M^{4+} - M^{5+}	
M^{5+} - M^{6+}	
M^{6+} - M^{7+}	
M^{7+} - M^{8+}	
M^{8+} - M^{9+}	
M^{9+} - M^{10+}	

Other Information

Enthalpy of Fusion/kJ mol^{-1} 17.2
Enthalpy of Vaporisation/kJ mol^{-1} 293

Oxidation States

main   Dy^{III}
others   Dy^{II}, Dy^{IV}

Covalent Bonds /kJ mol^{-1}

not applicable

Einsteinium   *Es*

General Information

Discovery

Einsteinium was discovered by G.R. Choppin, S.G. Thompson, A Ghiorso and B.G. Harvey in 1952, in the debris of the thermonuclear explosion in the Pacific at Eniwetok. This involved the examination of tons of radioactive coral from the blast area.

Appearance

Einsteinium is a radioactive, silvery metal.

Source

Einsteinium can be obtained in milligram quantities from the neutron bombardment of plutonium.

Uses

Einsteinium has no uses outside research.

Biological Role

Einsteinium has no known biological role. It is toxic due to its radioactivity.

General Information

Einsteinium is attacked by oxygen, steam and acids but not by alkalis.

Physical Information

Atomic Number	99
Relative Atomic Mass ($^{12}C=12.000$)	254 (radioactive)
Melting Point/K	not available
Boiling Point/K	not available
Density/kg m^{-3}	not available
Ground State Electron Configuration	$[Rn]5f^{11}7s^2$
Electron Affinity(M-M$^-$)/kJ mol^{-1}	50

Key Isotopes

nuclide	^{253}Es	^{254}Es
atomic mass		254.09
natural abundance	0%	0%
half-life	20.7 days	207 days

Ionisation Energies/kJ mol^{-1}

M - M$^+$	619
M$^+$ - M^{2+}	
M^{2+} - M^{3+}	
M^{3+} - M^{4+}	
M^{4+} - M^{5+}	
M^{5+} - M^{6+}	
M^{6+} - M^{7+}	
M^{7+} - M^{8+}	
M^{8+} - M^{9+}	
M^{9+} - M^{10+}	

Other Information

Enthalpy of Fusion/kJ mol^{-1}

not available

Enthalpy of Vaporisation/kJ mol^{-1}

not available

Oxidation States

main	Es^{III}
others	Es^{II}

Covalent Bonds /kJ mol^{-1}

not applicable

Erbium   *Er*

General Information

Discovery

Erbium was discovered by C.G. Mosander in 1842 in Stockholm, Sweden. It was first produced in reasonably pure form in 1934 by Klemm and Bonner.

Appearance

Erbium is a silver-grey metal, and is soft and malleable.

Source

Erbium is found principally in the minerals monazite and bastnasite, from which it can be extracted by ion exchange and solvent extraction.

Uses

Erbium is occasionally used in infra-red absorbing glass. Added to vanadium, it lowers the hardness and improves the workability. Otherwise it is little used.

Biological Role

Erbium has no known biological role, and has low toxicity.

General Information

Erbium slowly tarnishes in air, reacts slowly with water and dissolves in acids.

Physical Information

Atomic Number	68
Relative Atomic Mass ($^{12}C=12.000$)	167.26
Melting Point/K	1802
Boiling Point/K	3136
Density/kg m^{-3}	9066 (298K)
Ground State Electron Configuration	[Xe]4f^{12}6s^2
Electron Affinity(M-M$^-$)/kJ mol^{-1}	50

Key Isotopes

nuclide	^{162}Er	^{164}Er	^{166}Er	^{167}Er
atomic mass	161.9	163.9	165.9	166.9
natural abundance	0.14%	1.56%	33.4%	22.9%
half-life	stable	stable	stable	stable

nuclide	^{168}Er	^{169}Er	^{170}Er	^{171}Er
atomic mass	167.9		169.9	
natural abundance	27.1%	0%	14.9%	0%
half-life	stable	9.4 days	stable	7.52 h

Ionisation Energies/kJ mol^{-1}

M − M$^+$	588.7
M$^+$ − M^{2+}	1151
M^{2+} − M^{3+}	2194
M^{3+} − M^{4+}	4115
M^{4+} − M^{5+}	
M^{5+} − M^{6+}	
M^{6+} − M^{7+}	
M^{7+} − M^{8+}	
M^{8+} − M^{9+}	
M^{9+} − M^{10+}	

Other Information

Enthalpy of Fusion/kJ mol^{-1}	17.2
Enthalpy of Vaporisation/kJ mol^{-1}	280

Oxidation State Er^{III}

Covalent Bonds /kJ mol^{-1}

not applicable

Europium   *Eu*

General Information

Discovery

Europium was discovered by E.A. Demarcay in 1901 in Paris, France. The pure metal has only recently been prepared.

Appearance

Europium is a soft, silvery-white metal.

Source

In common with other rare earth elements, europium is found principally in the minerals monazite and basnaesite, from which it can be prepared. However, the usual method of preparation is by heating europium (III) oxide with an excess of lanthanum under vacuum.

Uses

Europium can absorb more neutrons per atom than any other element, making it valuable in control rods for nuclear reactors. Europium-doped plastic has been used as a laser material. Otherwise this element is very little used.

Biological Role

Europium has no known biological role, and has low toxicity.

General Information

Europium is the costliest and one of the rarest of the rare earth elements. It is as hard as lead and ductile, and is the most reactive of the rare earth metals, reacting rapidly with water and air.

Physical information

Atomic Number	63
Relative Atomic Mass ($^{12}C=12.000$)	151.97
Melting Point/K	1095
Boiling Point/K	1870
Density/kg m^{-3}	5243 (293K)
Ground State Electron Configuration	[Xe]4f^{7}6s^2
Electron Affinity(M-M$^-$)/kJ mol^{-1}	50

Key Isotopes

nuclide	^{151}Eu	^{152}Eu	^{153}Eu
atomic mass	150.9		152.9
natural abundance	47.8%	0%	52.2%
half-life	stable	12.7 yrs	stable

Ionisation Energies/kJ mol^{-1}

M − M$^+$	546.7
M$^+$ − M^{2+}	1085
M^{2+} − M^{3+}	2404
M^{3+} − M^{4+}	4110
M^{4+} − M^{5+}	
M^{5+} − M^{6+}	
M^{6+} − M^{7+}	
M^{7+} − M^{8+}	
M^{8+} − M^{9+}	
M^{9+} − M^{10+}	

Other Information

Enthalpy of Fusion/kJ mol^{-1} 10.5
Enthalpy of Vaporisation/kJ mol^{-1} 176

Oxidation States

main	Eu^{III}
others	Eu^{II}

Covalent Bonds /kJ mol^{-1}

not applicable

Fermium *Fm*

General Information

Discovery

Fermium was discovered by G. R. Choppin, S.G. Thompson, A. Ghioros and B.G. Harvey in 1952, in the debris of the thermonuclear explosion at Eniwetok in the Pacific. This involved the examination of tons of radioactive coral from the blast area.

Appearance

Fermium has a very short life-span, so scientists doubt that enough of the element will ever be obtained to be weighed or seen.

Source

Fermium can be obtained in microgram quantities from the neutron bombardment of plutonium.

Uses

Fermium has no uses outside research.

Biological Role

Fermium has no known biological role. It is toxic due to its radioactivity.

Physical information

Atomic Number	100
Relative Atomic Mass ($^{12}C=12.000$)	257 (radioactive)
Melting Point/K	not available
Boiling Point/K	not available
Density/kg m^{-3}	not available
Ground State Electron Configuration	[Rn]5f^{12}7s^2
Electron Affinity(M-M$^-$)/kJ mol^{-1}	not available

Key Isotopes

nuclide	^{254}Fm	^{255}Fm	^{257}Fm
atomic mass			
natural abundance	0%	0%	0%
half-life	3.24 h	20 h	80 days

Ionisation Energies/kJ mol^{-1}

M − M$^+$	627
M$^+$ − M^{2+}	
M^{2+} − M^{3+}	
M^{3+} − M^{4+}	
M^{4+} − M^{5+}	
M^{5+} − M^{6+}	
M^{6+} − M^{7+}	
M^{7+} − M^{8+}	
M^{8+} − M^{9+}	
M^{9+} − M^{10+}	

Other Information

Enthalpy of Fusion/kJ mol^{-1}
not available

Enthalpy of Vaporisation/kJ mol^{-1}
not available

Oxidation States Fm^{II}, Fm^{III}

Covalent Bonds /kJ mol^{-1}
not applicable

Fluorine   F

General Information

Discovery

Fluorine was first isolated by H. Moissan in Paris in 1886, after nearly 74 years of continuous effort by several investigators including Davy, Gay-Lussac, Lavoisier and Thenard.

Appearance

Fluorine is a pale yellow, corrosive gas. It has a characteristic pungent odour, detectable at very low concentrations.

Source

Fluorine occurs chiefly in the minerals fluorspar and cryolite, but is rather widely distributed in other minerals. It can be obtained by electrolysing a solution of potassium hydrogen fluoride in anhydrous hydrogen fluoride in a vessel of metal or transparent fluorspar.

Uses

There was no commercial production of fluorine until World War II, when the production of the atom bomb and other nuclear energy projects made it necessary to produce large quantities. The element and its compounds are used in producing uranium and many fluorochemicals, including high-temperature plastics. Hydrofluoric acid is used for etching the glass of light bulbs and similar applications, and fluorochloro-hydrocarbons are used in air conditioning and refrigeration. The presence of fluorides below 2ppm in drinking water is believed to prevent dental cavities, but above this concentration may cause mottled enamel in children acquiring permanent teeth.

Biological Role

The element fluorine and the fluoride ion are highly toxic.

General Information

Fluorine is the most reactive of the non-metals, and will combine with most other elements. Only a few of the inert gases do not combine with this element. It corrodes platinum, a metal that resists most other chemicals. In a stream of fluorine gas many substances burn with a bright flame, including finely-divided metals, glass, ceramics, carbon, wood, rubber and even water.

Physical Information

Atomic Number	9
Relative Atomic Mass ($^{12}C=12.000$)	19.998
Melting Point/K	53.53
Boiling Point/K	85.01
Density/kg m^{-3}	1.696 (gas, 273K)
Ground State Electron Configuration	[He]$2s^2 2p^5$
Electron Affinity(M-M$^-$)/kJ mol^{-1}	333

Key Isotopes

nuclide	^{18}F	^{19}F
atomic mass		18.998
natural abundance	0%	100%
half-life	109.7 mins	stable

Ionisation Energies/kJ mol^{-1}

M - M$^+$	1681
M$^+$ - M^{2+}	3374
M^{2+} - M^{3+}	6050
M^{3+} - M^{4+}	8408
M^{4+} - M^{5+}	11023
M^{5+} - M^{6+}	15164
M^{6+} - M^{7+}	17867
M^{7+} - M^{8+}	92036
M^{8+} - M^{9+}	106432

Other Information

Enthalpy of Fusion/kJ mol^{-1}	1.02
Enthalpy of Vaporisation/kJ mol^{-1}	3.26

Oxidation State F^{-I}

Covalent Bonds /kJ mol^{-1}

F-F	159
F-O	190
F-N	272

Francium *Fr*

General Information

Discovery

Francium was discovered by Marguerite Perey in 1939 at the Curie Institute, Paris.

Appearance

Francium has never actually been seen, as it is a short-lived product of the decay of actinium. It is a highly radioactive metal.

Source

Francium occurs as a result of the alpha disintegration of actinium, which is obtained from the neutron bombardment of radium. It can also be made artificially by bombarding thorium with protons.

Uses

Francium has no uses.

Biological Role

Francium has no known biological role. It is toxic due to its radioactivity.

General Information

Francium occurs naturally in uranium minerals but to an extremely small extent - there is probably less than 10g of francium at any one time in the crust of the earth. It is the most unstable of the first 101 elements of the Periodic Table. All its isotopes are highly unstable, so knowledge of its chemical properties comes from radiochemical techniques, and it most closely resembles caesium.

Physical Information

Atomic Number	87
Relative Atomic Mass ($^{12}C=12.000$)	223 (radioactive)
Melting Point/K	300
Boiling Point/K	950
Ground State Electron Configuration	$[Rn]7s^1$
Electron Affinity(M-M⁻)/kJ mol^{-1}	44

Key Isotopes

nuclide	^{212}Fr	^{223}Fr
atomic mass		223.02
natural abundance	0%	some
half-life	19 mins	22 mins

Ionisation Energies/kJ mol^{-1}

M − M$^+$	400
M$^+$ − M^{2+}	2100
M^{2+} − M^{3+}	3100
M^{3+} − M^{4+}	4100
M^{4+} − M^{5+}	5700
M^{5+} − M^{6+}	6900
M^{6+} − M^{7+}	8100
M^{7+} − M^{8+}	12300
M^{8+} − M^{9+}	12800
M^{9+} − M^{10+}	29300

Other Information

Oxidation State Fr^{I}

Covalent Bonds /kJ mol^{-1}

not applicable

Gadolinium   *Gd*

General Information

Discovery

Gadolinium was discovered by J.C. Galissard de Marignac in 1880 in Geneva, Switzerland. Lecoq de Boisbaudran isolated the element in 1886.

Appearance

Gadolinium is a silvery-white metal with a lustrous sheen.

Source

In common with other rare earth elements, gadolinium is found principally in the minerals monazite and basnaesite, from which it can be commercially prepared by ion exchange and solvent extraction. It is also prepared by reduction of the anhydrous fluoride with calcium metal.

Uses

Gadolinium has useful properties in alloys. As little as 1% gadolinium has been found to improve the workability and resistance of iron and chromium alloys to high temperatures and oxidation. It has found limited use in electronics, but is not a widely used metal.

Biological Role

Gadolinium has no known biological role, and has low toxicity.

General Information

Gadolinium reacts slowly with oxygen and water, and dissolves in acids. It is realtively stable in dry air but tarnishes in moist air.

Physical Information

Atomic Number	64
Relative Atomic Mass ($^{12}C=12.000$)	157.25
Melting Point/K	1586
Boiling Point/K	3539
Density/kg m^{-3}	7900 (298K)
Ground State Electron Configuration	[Xe]$4f^7 5d^1 6s^2$
Electron Affinity(M-M$^-$)/kJ mol^{-1}	50

Key Isotopes

nuclide	^{152}Gd	^{153}Gd	^{154}Gd	^{155}Gd
atomic mass	151.9		153.9	154.9
natural abundance	0.2%	0%	2.1%	14.8%
half-life	1.1×10^{14} yrs	242 days	stable	stable

nuclide	^{156}Gd	^{157}Gd	^{158}Gd	^{160}Gd
atomic mass	155.9	156.9	157.9	159.9
natural abundance	20.6%	15.6%	24.8%	21.9%
half-life	stable	stable	stable	stable

Ionisation Energies/kJ mol^{-1}

M - M$^+$	592.5
M$^+$ - M^{2+}	1167
M^{2+} - M^{3+}	1990
M^{3+} - M^{4+}	4250
M^{4+} - M^{5+}	
M^{5+} - M^{6+}	
M^{6+} - M^{7+}	
M^{7+} - M^{8+}	
M^{8+} - M^{9+}	
M^{9+} - M^{10+}	

Other Information

Enthalpy of Fusion/kJ mol^{-1} 15.5
Enthalpy of Vaporisation/kJ mol^{-1} 301

Oxidation States   Gd^{II}, Gd^{III}

Covalent Bonds /kJ mol^{-1}

not applicable

Gallium   *Ga*

General Information

Discovery

Gallium was discovered by P. Lecoq de Boisbaudran in 1875 in Paris. Mendeleev predicted and described this element, and called it ekaaluminum.

Appearance

Gallium is a silvery, glass-like, soft metal.

Source

Gallium is present in trace amounts in the minerals diaspore, sphalerite, germanite, bauxite and coal. The free metal can be obtained by electrolysis of a solution of gallium (III) hydroxide in potassium hydroxide.

Uses

Gallium readily alloys with most metals, and is used especially in low-melting alloys. It has a high boiling point which makes it ideal for recording temperatures that would vaporise a thermometer. It has found recent use in doping semiconductors and producing solid-state devices such as transistors.

Biological Role

Gallium has no known biological role. It is non-toxic.

General Information

Gallium is soluble in acids and alkalis. It has the longest liquid range of all elements, and can be liquid near room temperatures - it can melt in the hand. It also expands as it freezes, which is unusual for a metal, by 3.1%. Gallium wets glass or porcelain, and forms a brilliant mirror when painted on glass.

Physical Information

Atomic Number	31
Relative Atomic Mass (^{12}C=12.000)	69.723
Melting Point/K	302.9
Boiling Point/K	2676
Density/kg m^{-3}	5907 (293K)
Ground State Electron Configuration	$[Ar]3d^{10}4s^2 4p^1$
Electron Affinity(M-M$^-$)/kJ mol^{-1}	36

Key Isotopes

nuclide	^{67}Ga	^{69}Ga	^{71}Ga	^{72}Ga
atomic mass		68.926	70.925	
natural abundance	0%	60.1%	39.9%	0%
half-life	78.1 h	stable	stable	14.1 h

Ionisation Energies/kJ mol^{-1}

M – M$^+$	578.8
M$^+$ – M^{2+}	1979
M^{2+} – M^{3+}	2963
M^{3+} – M^{4+}	6200
M^{4+} – M^{5+}	8700
M^{5+} – M^{6+}	11400
M^{6+} – M^{7+}	14400
M^{7+} – M^{8+}	17700
M^{8+} – M^{9+}	22300
M^{9+} – M^{10+}	26100

Other Information

Enthalpy of Fusion/kJ mol^{-1} 5.59
Enthalpy of Vaporisation/kJ mol^{-1} 270.3

Oxidation States

main	Ga^{III}
others	Ga^{I}, Ga^{II}

Covalent Bonds /kJ mol^{-1}

not applicable

Germanium   *Ge*

General Information

Discovery

Germanium was discovered by C.A. Winkler in 1886 in Freiberg, Germany. It was predicted by Mendeleev in 1871 who named it ekasilicon.

Appearance

Germanium is a grey-white metalloid, crystalline and brittle, retaining a lustre in air.

Source

Germanium is found in small quantities in the minerals germanite and argyrodite. It is also present in zinc ores, and commercial production of germanium is by processing zinc smelter flue dust. It can also be recovered from the by-products of combustion of certain coals, which ensures a copious future supply.

Uses

Germanium is a very important semiconductor. The pure element is doped with arsenic, gallium or other elements and used as a transistor in thousands of electronic applications.

Germanium is also finding use as an alloying agent, in fluorescent lamps and as a catalyst. Both germanium and germanium oxide are transparent to infrared radiation and so are used in infrared spectroscopes. Germanium oxide has a high index of refraction and dispersion and is used in wide-angle camera lenses and microscope objectives.

Biological Role

Germanium has no known biological role. It is non-toxic. Certain germanium compounds have low mammalian toxicity but marked activity against some bacteria, which has stimulated interest in their use in pharmaceutical products.

Physical Information

Atomic Number	32
Relative Atomic Mass ($^{12}C=12.000$)	72.61
Melting Point/K	1210.6
Boiling Point/K	3103
Density/kg m^{-3}	5323 (293K)
Ground State Electron Configuration	$[Ar]3d^{10}4s^24p^2$
Electron Affinity(M-M$^-$)/kJ mol^{-1}	116

Key Isotopes

nuclide	^{68}Ge	^{70}Ge	^{71}Ge	^{72}Ge
atomic mass	67.928	69.924	70.925	71.923
natural abundance	0%	20.5%	0%	27.4%
half-life	287 days	stable	11.4 days	stable

nuclide	^{73}Ge	^{74}Ge	^{76}Ge	^{77}Ge
atomic mass	72.923	73.922	75.921	
natural abundance	7.8%	36.5%	7.8%	0%
half-life	stable	stable	stable	11.3 h

Ionisation Energies/kJ mol^{-1}

M − M$^+$	762.1
M$^+$ − M^{2+}	1537
M^{2+} − M^{3+}	3302
M^{3+} − M^{4+}	4410
M^{4+} − M^{5+}	9020
M^{5+} − M^{6+}	11900
M^{6+} − M^{7+}	15000
M^{7+} − M^{8+}	18200
M^{8+} − M^{9+}	21800
M^{9+} − M^{10+}	27000

Other Information

Enthalpy of Fusion/kJ mol^{-1}	34.7
Enthalpy of Vaporisation/kJ mol^{-1}	327.6

Oxidation States   Ge^{II}, Ge^{IV}

Covalent Bonds /kJ mol^{-1}

Ge-H	288
Ge-C	237
Ge-O	363
Ge-F	464
Ge-Cl	340
Ge-Ge	163

Gold   *Au*

General Information

Discovery

Gold was known to ancient civilisations, and has always been a valued metal.

Appearance

Of all the elements, gold is the most beautiful. It is a soft metal with a characteristic yellow colour and sheen.

Source

Gold is found in nature both free in veins and combined in alluvial deposits. About two thirds of the world's output comes from South Africa. Refining is usually by electrolysis, but gold in ores is recovered by a smelting process.

Uses

Gold is used for coinage and is a standard for monetary systems in many countries. It is also used extensively in jewellery. The term carat expresses the amount of gold present in an alloy; 24 carat is pure gold, and most jewellery is 9 carat gold. Gold is used in dental work, and the isotope ^{198}Au, with a half-life of 2.7 days, is used for treating cancer. A gold compound is used in certain cases to treat arthritis. Another gold compound is used in photography for toning the silver image.

Biological Role

Gold has no known biological role, and is non-toxic.

General Information

Gold has the highest malleability and ductility of any element. It is unaffected by air, water, all acids except aqua regia, and alkalis. It is a good conductor of heat and electricity. It is also a good reflector of infra-red radiation, and as it is inert makes an excellent coating for space satellites.

Physical Information

Atomic Number	79
Relative Atomic Mass ($^{12}C=12.000$)	196.97
Melting Point/K	1337.58
Boiling Point/K	3080
Density/kg m^{-3}	19320 (293K)
Ground State Electron Configuration	[Xe]4f^{14}5d^{10}6s^1
Electron Affinity(M-M$^-$)/kJ mol^{-1}	223

Key Isotopes

nuclide	^{195}Au	^{197}Au	^{198}Au	^{199}Au
atomic mass		196.97		
natural abundance	0%	100%	0%	0%
half-life	183 days	stable	2.69 days	3.15 days

Ionisation Energies/kJ mol^{-1}

M - M$^+$	890.1
M$^+$ - M^{2+}	1980
M^{2+} - M^{3+}	2900
M^{3+} - M^{4+}	4200
M^{4+} - M^{5+}	5600
M^{5+} - M^{6+}	7000
M^{6+} - M^{7+}	9300
M^{7+} - M^{8+}	11000
M^{8+} - M^{9+}	12800
M^{9+} - M^{10+}	14800

Other Information

Enthalpy of Fusion/kJ mol^{-1}	12.7
Enthalpy of Vaporisation/kJ mol^{-1}	343.1

Oxidation States

main   Au^{III}

others   Au^{-I}, Au^{O}, Au^{I}, Au^{II}, Au^{V}, Au^{VII}

Covalent Bonds /kJ mol^{-1}

not applicable

Hafnium — Hf

General Information

Discovery

Hafnium was discovered by D. Coster and G.C. von Hevesey in 1923 in Copenhagen, Denmark.

Appearance

Hafnium is a lustrous, silvery, ductile metal.

Source

Most zirconium minerals contain 1-5% hafnium, and the metal is prepared by reducing the tetrachloride with sodium or magnesium.

Uses

Hafnium has a good thermal absorption cross-section for neutrons, so is used in control rods in nuclear reactors. It has been successfully alloyed with several metals including iron, titanium and niobium. It is also used in gas-filled and incandescent lights.

Biological Role

Hafnium has no known biological role, and is non-toxic.

General Information

Hafnium resists corrosion due to an oxide film, but powdered hafnium will burn in air. It is unaffected by all acids except hydrogen fluoride, and also all alkalis. At high temperatures it reacts with oxygen, nitrogen, carbon, boron, sulphur and silicon.

Physical Information

Atomic Number	72
Relative Atomic Mass ($^{12}C=12.000$)	178.49
Melting Point/K	2503
Boiling Point/K	5470
Density/kg m^{-3}	13310 (293K)
Ground State Electron Configuration	$[Xe]4f^{14}5d^{2}6s^{2}$
Electron Affinity(M-M$^-$)/kJ mol^{-1}	-61

Key Isotopes

nuclide	^{172}Hf	^{174}Hf	^{175}Hf	^{176}Hf	^{177}Hf
atomic mass		173.9		175.9	176.9
natural abundance	0%	0.2%	0%	5.2%	18.6%
half-life	5 yrs	2×10^{15} yrs	70 days	stable	stable

nuclide	^{178}Hf	^{179}Hf	^{180}Hf	^{181}Hf	^{182}Hf
atomic mass	177.9	178.9	179.9		
natural abundance	27.1%	13.7%	35.2%	0%	0%
half-life	stable	stable	stable	42.5 days	9×10^{6} yrs

Ionisation Energies/kJ mol^{-1}

M - M^{+}	642
M^{+} - M^{2+}	1440
M^{2+} - M^{3+}	2250
M^{3+} - M^{4+}	3216
M^{4+} - M^{5+}	
M^{5+} - M^{6+}	
M^{6+} - M^{7+}	
M^{7+} - M^{8+}	
M^{8+} - M^{9+}	
M^{9+} - M^{10+}	

Other information

Enthalpy of Fusion/kJ mol^{-1} 25.5
Enthalpy of Vaporisation/kJ mol^{-1} 570.7

Oxidation States

main   Hf^{IV}
others   Hf^{I}, Hf^{II}, Hf^{III}

Covalent Bonds /kJ mol^{-1}

not applicable

Helium   He

General Information

Discovery

Helium was first detected by Janssen in 1868 during the solar eclipse as a new line in the solar spectrum, and named by Lockyer and Frankland. It was discovered in the uranium mineral clevite independently by Ramsay and the Swedish chemists Cleve and Langlet.

Appearance

Helium is a colourless gas, lighter than air.

Source

After hydrogen, helium is the second most abundant element in the universe. It has been detected spectroscopically in great abundance, especially in the hotter stars. It is present in the Earth's atmosphere in about 1 part in 200,000. It is present in various radioactive minerals as a decay product, but the major sources are from wells in Texas, Oklahoma and Kansas.

Uses

Helium is widely used as an inert gas shield for arc welding; as a protective gas in growing silicon and germanium crystals, and in titanium and zirconium production. It is also used as a cooling medium for nuclear reactors, and as a gas for supersonic wind tunnels. A mixture of 80% helium and 20% oxygen is used as an artificial atmosphere for divers and others working under pressure. Helium is extensively used for filling balloons as it is a much safer gas than hydrogen. One of the recent largest uses for helium has been for pressurising liquid fuel rockets.

Biological Role

Helium has no known biological function, but it is non-toxic.

General Information

Helium has the lowest melting point of any element and has found wide use in cryogenic research, as its boiling point is close to absolute zero. Its use in the study of superconductivity is vital.

Liquid helium (^{4}He) exists in two forms, ^{4}He I and ^{4}He II, above and below 2.174K respectively. The latter is unlike any other known substance. It expands on cooling, its conductivity for heat is enormous and neither its heat conduction nor viscocity obeys normal rules. It remains liquid down to absolute zero at ordinary pressures, but can readily be solidified by increasing the pressure.

Helium has a weak tendency to combine with other elements, for example in helium difluoride.

Physical Information

Atomic Number	2
Relative Atomic Mass ($^{12}C=12.000$)	4.003
Melting Point/K	0.95
Boiling Point/K	4.216
Density/kg m^{-3}	0.179 (gas, 273K)
Ground State Electron Configuration	$1s^2$
Electron Affinity(M-M$^-$)/kJ mol^{-1}	-21

Key Isotopes

nuclide	^{3}He	^{4}He
atomic mass	3.016	4.003
natural abundance	1.38x10^{-4}%	99.999%
half-life	stable	stable

Ionisation Energies/kJ mol^{-1}

M - M$^+$	2372.3
M$^+$ - M^{2+}	5250.4

Other Information

Enthalpy of Fusion/kJ mol^{-1} 0.021
Enthalpy of Vaporisation/kJ mol^{-1} 0.082

Oxidation States

not applicable

Covalent Bonds /kJ mol^{-1}

not applicable

Holmium　　　　　　　　　　*Ho*

General Information

Discovery

The spectral absorption bands of holmium were first identified by M. Delafontaine and J.L. Soret in 1878 in Geneva, Switzerland. The element was independently discovered by P.T. Cleve in 1878 in Uppsala, Sweden.

Appearance

Holmium is a silvery metal with a bright lustre.

Source

The principal source of holmium is the mineral monazite, from which it is obtained by ion exchange and solvent extraction. It can also be obtained by reduction of the anhydrous fluoride by calcium metal.

Uses

Holmium can absorb fission-bred neutrons, so is used in nuclear reactors as a burnable poison - it burns up while it is keeping a chain reaction under control. It is little used otherwise.

Biological Role

Holmium has no known biological role, and is non-toxic.

General Information

Holmium is relatively soft and malleable. It is slowly attacked by water and oxygen, and dissolves in acid.

Physical Information

Atomic Number	67
Relative Atomic Mass ($^{12}C=12.000$)	164.93
Melting Point/K	1747
Boiling Point/K	2968
Density/kg m^{-3}	8795 (298K)
Ground State Electron Configuration	[Xe]4f^{11}6s^2
Electron Affinity(M-M$^-$)/kJ mol^{-1}	50

Key Isotopes

nuclide	^{165}Ho	^{166}Ho
atomic mass	164.9	
natural abundance	100%	0%
half-life	stable	26.9 h

Ionisation Energies/kJ mol^{-1}

M – M$^+$	580.7
M$^+$ – M^{2+}	1139
M^{2+} – M^{3+}	2204
M^{3+} – M^{4+}	4100
M^{4+} – M^{5+}	
M^{5+} – M^{6+}	
M^{6+} – M^{7+}	
M^{7+} – M^{8+}	
M^{8+} – M^{9+}	
M^{9+} – M^{10+}	

Other Information

Enthalpy of Fusion/kJ mol^{-1} 17.2
Enthalpy of Vaporisation/kJ mol^{-1} 303

Oxidation State   Ho^{III}

Covalent Bonds /kJ mol^{-1}

not applicable

Hydrogen H

General Information

Discovery

Hydrogen was first recognized as an element by Cavendish in 1766, and named by Lavoisier.

Appearance

Hydrogen is a colourless gas.

Source

Hydrogen is found in the sun and most of the stars, and is easily the most abundant element in the universe. The planet Jupiter is composed mostly of hydrogen, and there is a theory that in the interior of the planet the pressure is so great that metallic hydrogen is formed from solid molecular hydrogen. On this planet, hydrogen is found in the greatest quantities in water, but is present in the atmosphere only in small amounts - less than 1 part per million by volume.

Hydrogen is prepared commercially by several methods; electrolysis of water, decomposition of hydrocarbons, displacement from acids by certain metals, action of steam on heated carbon, and action of sodium or potassium hydroxide on aluminium.

Uses

Large quantities are used in the Haber Process (the production of ammonia for agricultural use) and for the hydrogenation of fats and oils. It has several other uses, including welding and the reduction of metallic ores, and liquid hydrogen is important in cryogenics and superconductivity studies as its melting point is just above absolute zero.

Biological Role

Hydrogen is the basis of all life, as part of the DNA molecule.

General Information

There are three isotopes of hydrogen - protium, deuterium and tritium. Protium is the ordinary isotope, with an atomic mass of 1. Deuterium, atomic mass 2, was discovered in 1932 and tritium, atomic mass 3, in 1934. Tritium is unstable, with a half-life of about 12.5 years, and is used in nuclear reactors, hydrogen bombs, luminous paints and as a tracer. Protium is the most abundant isotope, and tritium the least abundant.

It would be possible to base the entire economy of the Earth on solar and nuclear generated hydrogen, an advantage as hydrogen itself is non-polluting, but the high cost of hydrogen compared to current hydrocarbon fuels makes this unrealistic at present.

Physical Information

Atomic Number	1
Relative Atomic Mass ($^{12}C=12.000$)	1.008
Melting Point/K	14.01
Boiling Point/K	20.28
Density/kg m^{-3}	0.09 (gas, 273K)
Ground State Electron Configuration	$1s^1$
Electron Affinity(M-M$^-$)/kJ mol^{-1}	72.8

Key Isotopes

nuclide	^1H	^2H	^3H
atomic mass	1.008	2.014	3.016
natural abundance	99.99%	0.015%	0%
half-life	stable	stable	12.262 yrs

Ionisation Energies/kJ mol^{-1}

M - M$^+$ 1312.0

Other Information

Enthalpy of Fusion/kJ mol^{-1} 0.12
Enthalpy of Vaporisation/kJ mol^{-1} 0.46

Oxidation States

main	H^I
others	H^0, H^{-I}

Covalent Bonds /kJ mol^{-1}

H-H	453.6
H-F	566
H-Cl	431
H-Br	366
H-I	299

Indium   *In*

General Information

Discovery

Indium was discovered by F. Reich and H. Richter in 1863 in Freiberg, Germany.

Appearance

Indium is a very soft, silvery-white metal with a brilliant lustre.

Source

Indium is often associated with zinc minerals and iron, lead and copper ores. It is commercially produced from the zinc minerals, usually as a by-product.

Uses

Indium has semiconductor uses in transistors, thermistors and photoconductors. It is also used to make low-temperature alloys; for example, an alloy of 24% indium-76% gallium is liquid at room temperature. Indium can also be plated onto metal and evaporated onto glass to give a mirror with better resistance than silver to corrosion. A tiny long-lived indium battery has been devised to power new electronic watches.

Biological Role

Indium has no known biological role but is teratogenic. It has a low order of toxicity.

General Information

Indium is stable in air and with water, but dissolves in acids.

Physical Information

Atomic Number	49
Relative Atomic Mass ($^{12}C=12.000$)	114.82
Melting Point/K	429.32
Boiling Point/K	2353
Density/kg m^{-3}	7310 (298K)
Ground State Electron Configuration	[Kr]$4d^{10}5s^{2}5p^{1}$
Electron Affinity(M-M$^-$)/kJ mol^{-1}	34

Key Isotopes

nuclide	^{111}In	^{113}In	^{115}In
atomic mass		112.9	114.9
natural abundance	0%	4.3%	95.7%
half-life	2.81 days	stable	6×10^{14} yrs

Ionisation Energies/kJ mol^{-1}

M − M$^+$	558.3
M$^+$ − M^{2+}	1820.6
M^{2+} − M^{3+}	2704
M^{3+} − M^{4+}	5200
M^{4+} − M^{5+}	7400
M^{5+} − M^{6+}	9500
M^{6+} − M^{7+}	11700
M^{7+} − M^{8+}	13900
M^{8+} − M^{9+}	17200
M^{9+} − M^{10+}	19700

Other Information

Enthalpy of Fusion/kJ mol^{-1}	3.27
Enthalpy of Vaporisation/kJ mol^{-1}	231.8

Oxidation States

main	In^{III}
others	In^{I}, In^{II}

Covalent Bonds /kJ mol^{-1}

not applicable

Iodine   I

General Information

Discovery

Iodine was discovered by B. Courtois in 1811 in Paris, France.

Appearance

Iodine is a blue-black, shiny solid which sublimes at room temperature into a blue-violet gas with an irritating odour.

Source

Iodine occurs sparingly (0.05 parts per million) in sea-water. From this source it is assimilated by seaweeds, brines from old sea deposits and brackish waters from oil and salt wells.

Iodine is obtained commercially by extracting iodine vapour from processed brine, by ion exchange of brine or by liberating iodine from iodate obtained from nitrate ores.

Uses

Iodine is used in several areas including pharmaceuticals, photographic chemicals, printing inks and dyes, catalysts and animal feeds.

Biological Role

Iodine is an essential element, lack of which causes problems with the thyroid gland. The artificial radioisotope, ^{131}I, with a half-life of 8 days, is used in treating the thyroid gland.

A solution of potassium iodide and iodine has germicidal effects and is used for the external treatment of wounds.

If iodine is in contact with the skin it can cause lesions, and iodine vapour is extremely irritating to the eyes and mucous membranes.

General Information

Iodine forms compounds with many elements, but is less active than the other halogens. It dissolves readily in chloroform, carbon tetrachloride and carbon disulphide to form beautiful purple solutions. It is only sparingly soluble in water. Organic iodine compounds are important in organic chemistry.

Physical Information

Atomic Number	53
Relative Atomic Mass ($^{12}C=12.000$)	126.9
Melting Point/K	386.7
Boiling Point/K	457.5
Density/kg m^{-3}	4930 (293K)
Ground State Electron Configuration	[Kr]4d^{10}5s^{2}5p^5
Electron Affinity(M-M$^-$)/kJ mol^{-1}	295

Key Isotopes

nuclide	^{123}I	^{125}I	^{127}I	^{129}I	^{131}I
atomic mass			126.9		
natural abundance	0%	0%	100%	0%	0%
half-life	13.3 h	60.2 days	stable	1.7x10^7 yrs	8 days

Ionisation Energies/kJ mol^{-1}

M - M$^+$	1008.4
M$^+$ - M^{2+}	1845.9
M^{2+} - M^{3+}	3200
M^{3+} - M^{4+}	4100
M^{4+} - M^{5+}	5000
M^{5+} - M^{6+}	7400
M^{6+} - M^{7+}	8700
M^{7+} - M^{8+}	16400
M^{8+} - M^{9+}	19300
M^{9+} - M^{10+}	22100

Other Information

Enthalpy of Fusion/kJ mol^{-1}	15.27
Enthalpy of Vaporisation/kJ mol^{-1}	41.67

Oxidation States

main	I^{-I}
others	I^O, I^{III}, I^V, I^{VII}

Covalent Bonds /kJ mol^{-1}

I-H	299
I-C	218
I-O	234
I-F	280
I-Cl	208
I-I	151
I-Si	234
I-P	184

Iridium   Ir

General Information

Discovery
Iridium was discovered by S. Tennant in 1803 in London.

Appearance
Iridium is a hard, lustrous, platinum-like metal.

Source
Iridium occurs uncombined in nature in alluvial deposits, and is recovered commercially as a by-product of nickel refining.

Uses
Iridium is used principally as a hardening agent for platinum. It also forms an alloy with osmium which is used for pen tips and compass bearings. It is the most corrosion-resistant material known, and was used in making the standard metre bar, which is an alloy of 90% platinum and 10% iridium.

Biological Role
Iridium has no known biological role, and has low toxicity.

Physical Information

Atomic Number	77
Relative Atomic Mass ($^{12}C=12.000$)	192.2
Melting Point/K	2683
Boiling Point/K	4403
Density/kg m^{-3}	22560 (290K)
Ground State Electron Configuration	[Xe]$4f^{14}5d^{7}6s^{2}$
Electron Affinity(M-M$^-$)/kJ mol^{-1}	190

Key Isotopes

nuclide	^{191}Ir	^{192}Ir	^{193}Ir
atomic mass	190.96		192.96
natural abundance	37.3%	0%	62.7%
half-life	stable	74.2 days	stable

Ionisation Energies/kJ mol^{-1}

M − M$^+$	880
M$^+$ − M^{2+}	1680
M^{2+} − M^{3+}	2600
M^{3+} − M^{4+}	3800
M^{4+} − M^{5+}	5500
M^{5+} − M^{6+}	6900
M^{6+} − M^{7+}	8500
M^{7+} − M^{8+}	10000
M^{8+} − M^{9+}	11700
M^{9+} − M^{10+}	

Other Information

Enthalpy of Fusion/kJ mol^{-1} 26.4
Enthalpy of Vaporisation/kJ mol^{-1} 612.1

Oxidation States

main   Ir^{III}, Ir^{IV}
others   Ir^{-I}, Ir^{O}, Ir^{I}, Ir^{II}, Ir^{V}, Ir^{VI}

Covalent Bonds /kJ mol^{-1}

not applicable

Iron   *Fe*

General Information

Discovery

The use of iron was known to ancient civilisations.

Appearance

Iron is a lustrous, silvery and soft metal. It can be worked relatively easily.

Source

Iron is the fourth most abundant element, by mass, in the crust of the earth. The core of the earth is thought to be largely composed of iron with about 10% occluded hydrogen. The commonest iron-containing ore is haematite, but iron is found widely distributed in other minerals such as magnetite and taconite.

Commercially, iron is produced by the reduction of haematite or magnetite with carbon in a furnace, the carbon being produced in situ by the burning of coke.

Uses

Iron is the most useful of all metals. It is also the cheapest available metal. The major proportion is used to manufacture steel, as carbon steel is an alloy of iron with carbon, with small amounts of other elements. Alloy steels are carbon steels with other additives such as nickel and chromium.

Wrought iron is iron containing a very small amount of carbon, and is tough, malleable and less fusible than pure iron.

Pig iron is an alloy containing about 3% carbon with varying amounts of sulphur, silicon, manganese and phosphorus. It is hard, brittle, fairly fusible and is used to produce other alloys including steel.

Biological Role

Iron is an essential and non-toxic element. It is part of the active site of haemoglobin, and carries oxygen in the bloodstream. Insufficient iron in the blood is the cause of anaemia.

General Information

Pure iron is very reactive chemically and rapidly rusts, especially in moist air or high temperatures. It dissolved in dilute acids.

Iron exists as four allotropic forms, one of which is magnetic. The relationship between these forms is not properly understood.

A remarkable wrought iron pillar which dates from A.D. 400 still stands today in Delhi, India. It is 7.25m high and 40cm in diameter. Corrosion to the pillar has been minimal although it has been constantly exposed to the weather.

Physical Information

Atomic Number	26
Relative Atomic Mass ($^{12}C=12.000$)	55.847
Melting Point/K	1808
Boiling Point/K	3023
Density/kg m^{-3}	7874 (293K)
Ground State Electron Configuration	[Ar]$3d^6 4s^2$
Electron Affinity(M-M$^-$)/kJ mol^{-1}	44

Key Isotopes

nuclide	^{52}Fe	^{54}Fe	^{55}Fe	^{56}Fe
atomic mass		53.940	54.938	55.935
natural abundance	0%	5.8%	0%	91.2%
half-life	8.2 h	stable	2.6 yrs	stable

nuclide	^{57}Fe	^{58}Fe	^{59}Fe	^{60}Fe
atomic mass	56.935	57.933	58.935	
natural abundance	2.2%	0.3%	0%	0%
half-life	stable	stable	45.1 days	3×10^5 yrs

Ionisation Energies/kJ mol^{-1}

M - M$^+$	759.3
M$^+$ - M^{2+}	1561
M^{2+} - M^{3+}	2957
M^{3+} - M^{4+}	5290
M^{4+} - M^{5+}	7240
M^{5+} - M^{6+}	9600
M^{6+} - M^{7+}	12100
M^{7+} - M^{8+}	14575
M^{8+} - M^{9+}	22678
M^{9+} - M^{10+}	25290

Other Information

Enthalpy of Fusion/kJ mol^{-1}	14.9
Enthalpy of Vaporisation/kJ mol^{-1}	340.2

Oxidation States

main	Fe^{II}, Fe^{III}
others	Fe^{-II}, Fe^{-I}, Fe^{O}, Fe^{I}, Fe^{IV}, Fe^{V}, Fe^{VI}

Covalent Bonds /kJ mol^{-1}

not applicable

Krypton   *Kr*

General Information

Discovery

Krypton was discovered by Sir William Ramsey and M.W. Travers in 1898 in London, in the residue remaining after liquid air had boiled away.

Appearance

Krypton is a colourless, odourless gas.

Source

Krypton is obtained by distillation from liquid air.

Uses

Krypton is used commercially as a low-pressure filling gas for fluorescent lights. It is also used in certain photographic flash lamps for high-speed photography. Radioactive krypton is used to estimate Soviet nuclear production. The gas is a product of all nuclear reactors, so the Russian share is found by subtracting the amount that comes from Western reactors from the total in the air.

Biological Role

Krypton has no known biological role.

General Information

The spectral lines of krypton - brilliant green and orange - are easily produced and very sharp. The orange-red line of ^{86}Kr is used as the fundamental unit of length: 1 metre=1650763.73 wavelengths. Some krypton compounds can be made, including krypton (II) flouride and clathrates.

Physical Information

Atomic Number	36
Relative Atomic Mass (^{12}C=12.000)	83.8
Melting Point/K	116.6
Boiling Point/K	120.85
Density/kg m^{-3}	3.75 (gas, 273K)
Ground State Electron Configuration	[Ar]3d^{10}4s^{2}4p^{6}
Electron Affinity(M-M$^-$)/kJ mol^{-1}	-39

Key Isotopes

nuclide	^{78}Kr	^{80}Kr	^{82}Kr	^{83}Kr
atomic mass	77.92	79.92	81.91	82.91
natural abundance	0.35%	2.25%	11.6%	11.5%
half-life	stable	stable	stable	stable

nuclide	^{84}Kr	^{85}Kr	^{86}Kr
atomic mass	83.91	84.91	85.91
natural abundance	57.0%	0%	17.3%
half-life	stable	10.76 yrs	stable

Ionisation Energies/kJ mol^{-1}

M − M$^+$	1350.7
M$^+$ − M^{2+}	2350
M^{2+} − M^{3+}	3565
M^{3+} − M^{4+}	5070
M^{4+} − M^{5+}	6240
M^{5+} − M^{6+}	7570
M^{6+} − M^{7+}	10710
M^{7+} − M^{8+}	12200
M^{8+} − M^{9+}	22229
M^{9+} − M^{10+}	28900

Other Information

Enthalpy of Fusion/kJ mol^{-1}	1.64
Enthalpy of Vaporisation/kJ mol^{-1}	9.05

Oxidation States   Kr^0, Kr^{II}

Covalent Bonds /kJ mol^{-1}

not applicable

Lanthanum   *La*

General Information

Discovery

Lanthanum was discovered by C.G. Mosander in 1839 in Stockholm, Sweden.

Appearance

Lanthanum is a silvery-white metal which can be cut with a knife. It is ductile, malleable and tarnishes in air.

Source

Lanthanum is found in rare earth minerals, principally monazite (25% lanthanum) and bastnaesite (38% lanthanum). Ion-exchange and solvent extraction techniques enable the rare earth elements to be isolated from the mineral, and lanthanum is usually obtained by reducing the anhydrous fluoride with calcium.

Uses

Rare earth compounds containing lanthanum are used extensively in carbon lighting applications, such as studio lighting and cinema projection. Lanthanum (III) oxide is used in making special optical glasses, as it improves the alkali resistance of glass. The ion lanthanum^{3+} is used as a biological tracer for Ca^{2+}, and radioactive lanthanum has been tested for use in treating cancer.

Biological Role

Lanthanum has no known biological role, but both the element and its compounds are moderately toxic.

General Information

Lanthanum is one of the most reactive of the rare earth metals. It oxidises rapidly when exposed to the air, and burns easily. It is attacked slowly by cold water, and rapidly by hot water. It reacts directly with carbon, nitrogen, phosphorus, sulphur, the halogens and some other elements. At room temperature, the structure of lanthanum is hexagonal. This changes to face-centred cubic at 310K, and to body-centred cubic at 865K.

Physical Information

Atomic Number	57
Relative Atomic Mass ($^{12}C=12.000$)	138.91
Melting Point/K	1194
Boiling Point/K	3730
Density/kg m^{-3}	6145 (298K)
Ground State Electron Configuration	[Xe]$5d^1 6s^2$
Electron Affinity(M-M$^-$)/kJ mol^{-1}	53

Key Isotopes

nuclide	^{138}La	^{139}La	^{140}La
atomic mass	137.9	138.9	
natural abundance	0.09%	99.91%	0%
half-life	stable		40.22 h

Ionisation Energies/kJ mol^{-1}

M - M$^+$	538.1
M$^+$ - M^{2+}	1067
M^{2+} - M^{3+}	1850
M^{3+} - M^{4+}	4819
M^{4+} - M^{5+}	6400
M^{5+} - M^{6+}	7600
M^{6+} - M^{7+}	9600
M^{7+} - M^{8+}	11000
M^{8+} - M^{9+}	12400
M^{9+} - M^{10+}	15900

Other Information

Enthalpy of Fusion/kJ mol^{-1}	10.04
Enthalpy of Vaporisation/kJ mol^{-1}	402.1

Oxidation State   La^{III}

Covalent Bonds /kJ mol^{-1}

not applicable

Lawrencium   *Lr*

General Information

Discovery

Lawrencium was discovered by A. Ghiorso and co-workers in 1961 in California, USA.

Appearance

Lawrencium is a radioactive metal. Only a few atoms have ever been made, so its appearance is unknown.

Source

Lawrencium is produced by bombarding californium with boron nuclei.

Uses

Lawrencium has no uses outside research.

Biological Role

Lawrencium has no known biological role. It is toxic due to its radioactivity.

Physical Information

Atomic Number	103
Relative Atomic Mass ($^{12}C=12.000$)	260 (radioactive)
Melting Point/K	not available
Boiling Point/K	not available
Density/kg m^{-3}	not available
Ground State Electron Configuration	[Rn]$5f^{14}6d^{1}7s^{2}$
Electron Affinity(M-M$^-$)/kJ mol^{-1}	not available

Key Isotopes

nuclide	^{260}Lr
atomic mass	
natural abundance	0%
half-life	3 mins

Ionisation Energies/kJ mol^{-1}

M - M$^+$ not applicable
M$^+$ - M^{2+}
M^{2+} - M^{3+}
M^{3+} - M^{4+}
M^{4+} - M^{5+}
M^{5+} - M^{6+}
M^{6+} - M^{7+}
M^{7+} - M^{8+}
M^{8+} - M^{9+}
M^{9+} - M^{10+}

Other Information

Enthalpy of Fusion/kJ mol^{-1}
not available

Enthalpy of Vaporisation/kJ mol^{-1}
not available

Oxidation State   Lr^{III}

Covalent Bonds /kJ mol^{-1}
not applicable

Lead   *Pb*

General Information

Discovery

Lead was known to ancient civilisations, and is mentioned in Exodus.

Appearance

Lead is a soft, weak, ductile metal with a pale grey sheen.

Source

Lead is obtained chiefly from the mineral galena by a roasting process. At least 40% of lead in the UK comes from secondary lead sources such as scrap batteries and pipes.

Uses

Lead is very resistant to corrosion - lead pipes from Roman times are still in use today - and is often used to store corrosive liquids. Great quantities of lead, both as the metal and the dioxide, are used in batteries. Lead is also used in cable covering, plumbing and ammunition. Tetraethyl lead is used as an anti-knock agent in petrol, and as an additive in paints. The use of lead in plumbing, petrol and paints has been reduced in the past few years because of environmental concern, as lead is a cumulative poison and is thought to affect brain development and function, especially in young children. Lead is an effective shield around X-ray equipment and nuclear reactors. Lead oxide is used in the production of fine crystal glass.

Biological Role

Lead has no known biological role. It is toxic in a cumulative way, teratogenic and carcinogenic.

General Information

Lead is stable to air and water, but will tarnish in moist air over long periods. It dissolves in nitric acid. Lead is a poor conductor of electricity.

Physical Information

Atomic Number	82
Relative Atomic Mass ($^{12}C=12.000$)	207.2
Melting Point/K	600.65
Boiling Point/K	2013
Density/kg m^{-3}	11350 (293K)
Ground State Electron Configuration	[Xe]4f^{14}5d^{10}6s^{2}6p^2
Electron Affinity(M-M$^-$)/kJ mol^{-1}	35.2

Key Isotopes

nuclide	^{204}Pb	^{205}Pb	^{206}Pb	^{207}Pb	^{208}Pb
atomic mass	203.97		205.97	206.98	207.98
natural abundance	1.4%	0%	24.1%	22.1%	52.3%
half-life	stable	3x10^{7}yrs	stable	stable	stable

Ionisation Energies/kJ mol^{-1}

M - M$^+$	715.5
M$^+$ - M^{2+}	1450.4
M^{2+} - M^{3+}	3081.5
M^{3+} - M^{4+}	4083
M^{4+} - M^{5+}	6640
M^{5+} - M^{6+}	8100
M^{6+} - M^{7+}	9900
M^{7+} - M^{8+}	11800
M^{8+} - M^{9+}	13700
M^{9+} - M^{10+}	16700

Other Information

Enthalpy of Fusion/kJ mol^{-1}	5.12
Enthalpy of Vaporisation/kJ mol^{-1}	177.8

Oxidation States Pb^{II}, Pb^{IV}

Covalent Bonds /kJ mol^{-1}

Pb-H	180
Pb-C	130
Pb-O	398
Pb-F	314
Pb-Cl	244
Pb-Pb	100

Lithium   *Li*

General Information

Discovery

Lithium was discovered by Arfvedson in 1817.

Appearance

Lithium has a silvery appearance but quickly becomes covered by a film of black oxide when exposed to air. It is usually stored immersed in an inert oil.

Source

Lithium does not occur free in nature, but is found combined in small amounts in nearly all igneous rocks and in the waters of many mineral springs. Lepidolite, spodumene, petalite and amblygonite are the more important minerals containing lithium. Large deposits of spodumene are recovered from brines of lakes in California and Nevada, and solid deposits are found in North Carolina. Lithium metal is usually produced electrolytically from the fused chloride.

Uses

Lithium has the highest specific heat of any solid element, and is therefore used in many heat transfer applications. However, it is corrosive and requires special handling. It is used as an alloying agent, in the synthesis of organic compounds, and has nuclear applications. It has a high electrochemical potential so is one of the most widely used battery anode materials. Lithium is also used in special glasses and ceramics.

Lithium chloride is one of the most hygroscopic materials known, and is used in air conditioning and industrial drying systems (as is lithium bromide). Lithium stearate is used as an all-purpose and high-temperature lubricant.

Biological Role

Lithium has no known natural biological role. It is non-toxic, teratogenic, stimulatory and an anti-depressant.

General Information

Lithium reacts with water, but not as vigorously as sodium. It imparts a beautiful crimson colour to a flame, but when the metal burns strongly the flame is a dazzling white.

Physical Information

Atomic Number	3
Relative Atomic Mass ($^{12}C=12.000$)	6.941
Melting Point/K	453.69
Boiling Point/K	1620
Density/kg m^{-3}	534 (293K)
Ground State Electron Configuration	[He]2s^1
Electron Affinity(M-M$^-$)/kJ mol^{-1}	57

Key Isotopes

nuclide	^{6}Li	^{7}Li
atomic mass	6.015	7.016
natural abundance	7.5%	92.5%
half-life	stable	stable

Ionisation Energies/kJ mol^{-1}

M - M$^+$	513.3
M$^+$ - M^{2+}	7298.0
M^{2+} - M^{3+}	11814.8

Other Information

Enthalpy of Fusion/kJ mol^{-1}	4.60
Enthalpy of Vaporisation/kJ mol^{-1}	147.7

Oxidation States

main	Li^I
other	Li^{-I}

Covalent Bonds /kJ mol^{-1}

not applicable

Lutetium *Lu*

General Information

Discovery

Lutetium was discovered by G. Urbain in 1907 in Paris, France, and independently by C. James in the same year in New Hampshire, USA.

Appearance

Lutetium is a silvery-white metal, the hardest and densest of the rare earth elements.

Source

In common with many other rare earth elements, the principal source of lutetium is the mineral monazite, from which it is extracted with difficulty by reduction of the anhydrous fluoride by a metal from Group I or II.

Uses

Lutetium has no practical value.

Biological Role

Lutetium has no known biological role, and has low toxicity.

General Information

Lutetium is one of the costliest of the rare earth elements. It is relatively stable in air.

Physical Information

Atomic Number	71
Relative Atomic Mass ($^{12}C=12.000$)	174.97
Melting Point/K	1963
Boiling Point/K	3668
Density/kg m^{-3}	9840 (298K)
Ground State Electron Configuration	[Xe]$4f^{14}5d^{1}6s^{2}$
Electron Affinity(M-M$^-$)/kJ mol^{-1}	50

Key Isotopes

nuclide	^{175}Lu	^{176}Lu	^{177}Lu
atomic mass	174.9		
natural abundance	97.39%	2.61%	0%
half-life	stable	2.2×10^{10} yrs	6.74 days

Ionisation Energies/kJ mol^{-1}

M - M$^+$	523.5
M$^+$ - M^{2+}	1340
M^{2+} - M^{3+}	2022
M^{3+} - M^{4+}	4360
M^{4+} - M^{5+}	
M^{5+} - M^{6+}	
M^{6+} - M^{7+}	
M^{7+} - M^{8+}	
M^{8+} - M^{9+}	
M^{9+} - M^{10+}	

Other information

Enthalpy of Fusion/kJ mol^{-1} 19.2
Enthalpy of Vaporisation/kJ mol^{-1} 428

Oxidation State   Lu^{III}

Covalent Bonds /kJ mol^{-1}

not applicable

Magnesium   Mg

General Information

Discovery

Joseph Black recognised magnesium as an element in 1755, but it was first isolated by Sir Humphrey Davy in 1808, and prepared in coherent form by Bussy in 1831.

Appearance

Magnesium is a sivery-white, lustrous and relatively soft metal, which tarnishes slightly in air.

Source

Magnesium is the eighth most abundant element in the earth's crust, but does not occur uncombined. It is found in large deposits in minerals such as magnesite and dolomite. Commercially, it is prepared by electrolysis of fused magnesium chloride derived from brines, wells and sea water.

Uses

Magnesium is used in photography, flares, pyrotechnics and incendiary bombs. As it is one third lighter than aluminium, its alloys are useful in aeroplane and missile construction.

It improves the mechanical, fabrication and welding characteristics of aluminium when used as an alloying agent.

Magnesium hydoxide (milk of magnesia), sulphate (Epsom salts), chloride and citrate are used in medicine.

Grignard reagents, which are organic magnesium compounds, are important commercially.

Biological Role

Magnesium is an essential element in both plant and animal life. It is non-toxic. Chlorophyls are magnesium-centred porphyrins.

General Information

Great care should be taken in handling magnesium metal, especially in the finely-divided state, as serious fires can occur. Water should not be used on burning magnesium or magnesium fires.

Physical Information

Atomic Number	12
Relative Atomic Mass ($^{12}C=12.000$)	24.305
Melting Point/K	922.0
Boiling Point/K	1363
Density/kg m^{-3}	1738 (293K)
Ground State Electron Configuration	[Ne]3s^2
Electron Affinity(M-M$^-$)/kJmol^{-1}	-67

Key Isotopes

nuclide	^{24}Mg	^{25}Mg	^{26}Mg
atomic mass	23.985	24.986	25.983
natural abundance	78.99%	10.00%	11.01%
half-life	stable	stable	stable

Ionisation Energies/kJ mol^{-1}

M - M$^+$	737.7
M$^+$ - M^{2+}	1450.7
M^{2+} - M^{3+}	7732.6
M^{3+} - M^{4+}	10540
M^{4+} - M^{5+}	13630
M^{5+} - M^{6+}	17995
M^{6+} - M^{7+}	21703
M^{7+} - M^{8+}	25656
M^{8+} - M^{9+}	31642
M^{9+} - M^{10+}	35461

Other Information

Enthalpy of Fusion/kJ mol^{-1}	9.04
Enthalpy of Vaporisation/kJ mol^{-1}	127.6

Oxidation State   Mg^{II}

Covalent Bonds /kJ mol^{-1}

not applicable

Manganese   *Mn*

General Information

Discovery

Manganese was recognised as an element by Scheele, Bergman and others and isolated by J.G. Grahn in 1774 in Stockholm, Sweden.

Appearance

Manganese is a grey-white metal, resembling iron, but is harder and very brittle.

Source

Manganese minerals are widely distributed, pyrolusite and rhodochrosite being the most common. Manganese nodules have been found on the floor of the oceans. These nodules contain about 24% manganese together with many other elements in lesser abundance.

Uses

Manganese is used to form many important alloys. It gives steel a hard yet pliant quality, and with aluminium and antimony it forms highly ferromagnetic alloys.

Manganese (IV) oxide is used as a depolariser in dry cells, and to decolorise glass coloured green by iron impurities. Manganese (II) oxide is a powerful oxidising agent and is used in quantitative analysis and in medicine.

Biological Role

Manganese is an essential element. Without it, bones grow spongier and break more easily. It activates many enzymes and may be essential for utilization of vitamin B. Exposure to manganese dust, fumes and compounds is to be avoided as it is a suspected carcinogen.

General Information

Manganese is reactive chemically, and decomposes cold water slowly. It is reactive even when impure, and will burn in oxygen.

Physical Information

Atomic Number	25
Relative Atomic Mass ($^{12}C=12.000$)	54.938
Melting Point/K	1517
Boiling Point/K	2235
Density/kg m^{-3}	7440 (293K)
Ground State Electron Configuration	[Ar]$3d^5 4s^2$
Electron Affinity(M-M$^-$)/kJ mol^{-1}	-94

Key Isotopes

nuclide	^{53}Mn	^{54}Mn	^{55}Mn	^{56}Mn
atomic mass	52.941	53.940	54.938	
natural abundance	0%	0%	100%	0%
half-life	2×10^6 yrs	303 days	stable	2.576 h

Ionisation Energies/kJ mol^{-1}

M - M$^+$	717.4
M$^+$ - M^{2+}	1509.0
M^{2+} - M^{3+}	3248.4
M^{3+} - M^{4+}	4940
M^{4+} - M^{5+}	6990
M^{5+} - M^{6+}	9200
M^{6+} - M^{7+}	11508
M^{7+} - M^{8+}	18956
M^{8+} - M^{9+}	21400
M^{9+} - M^{10+}	23960

Other Information

Enthalpy of Fusion/kJ mol^{-1}	14.4
Enthalpy of Vaporisation/kJ mol^{-1}	220.5

Oxidation States

main: Mn^{II}

others: Mn^{-III}, Mn^{-II}, Mn^{-I}, Mn^{O}, Mn^{I}, Mn^{III}, Mn^{IV}, Mn^{V}, Mn^{VI}, Mn^{VII}

Covalent Bonds /kJ mol^{-1}

not applicable

Mendelevium *Md*

General Information

Discovery

Mendelevium was discovered by A. Ghiorso and co-workers in 1955 in California, USA.

Appearance

Mendelevium is a radioactive metal. Only a few atoms have ever been made so its appearance is unknown.

Source

Mendelevium is made by bombarding einsteinium with alpha-particles.

Uses

Mendelevium is used only for research.

Biological Role

Mendelevium has no known biological role. It is toxic due to its radioactivity.

Physical Information

Atomic Number	101
Relative Atomic Mass ($^{12}C=12.000$)	not available
Melting Point/K	not available
Boiling Point/K	not available
Density/kg m^{-3}	not available
Ground State Electron Configuration	[Rn]5f^{13}7s^2
Electron Affinity(M-M$^-$)/kJ mol^{-1}	not available

Key Isotopes

nuclide ^{258}Md

atomic mass

natural abundance 0%

half-life 54 days

Ionisation Energies/kJ mol^{-1}

M - M$^+$ 635

M$^+$ - M^{2+}

M^{2+} - M^{3+}

M^{3+} - M^{4+}

M^{4+} - M^{5+}

M^{5+} - M^{6+}

M^{6+} - M^{7+}

M^{7+} - M^{8+}

M^{8+} - M^{9+}

M^{9+} - M^{10+}

Other Information

Enthalpy of Fusion/kJ mol^{-1}

not available

Enthalpy of Vaporisation/kJ mol^{-1}

not available

Oxidation States

main Md^{III}

other Md^{II}

Covalent Bonds /kJ mol^{-1}

not applicable

Mercury — Hg

General Information

Discovery

Mercury was known to ancient civilisations, such as the Chinese and Hindus, and has been found in Egyptian tombs of 1500B.C.

Appearance

Mercury is a heavy, silvery, liquid metal.

Source

Mercury occurs very rarely free in nature, but can be found in ores, principally cinnabar. This is mostly found in Spain and Italy, which together produce about 50% of the world's supply of this element. The metal is obtained by heating cinnabar in a current of air and condensing the vapour.

Uses

Mercury easily forms alloys with other metals such as gold, silver and tin, which are called amalgams. Its ease in amalgamating with gold is made use of in recovering gold from its ores. The metal is widely used in making advertising signs, mercury switches and other electrical apparatus. It is used in laboratory work for making thermometers, barometers, diffusion pumps and many other instruments. Other uses are in pesticides, dental work, batteries and catalysts.

Some mercury salts, and organic mercury compounds are important.

Biological Role

Mercury has no known biological role. It is a virulent poison, readily absorbed through the respiratory tract, the gastrointestinal tract or through the skin. It is a cumulative poison and dangerous levels are readily attained in air. It is always handled with the utmost care.

General Information

Mercury is stable with air and water, unreactive to all acids except nitric acid, and all alkalis. It is a rather poor conductor of heat compared to other metals, and a fair conductor of electricity.

Physical Information

Atomic Number	80
Relative Atomic Mass ($^{12}C=12.000$)	200.59
Melting Point/K	234.28
Boiling Point/K	629.73
Density/kg m^{-3}	13546 (293K)
Ground State Electron Configuration	[Xe]4f^{14}5d^{10}6s^2
Electron Affinity(M-M$^-$)/kJ mol^{-1}	-18

Key Isotopes

nuclide	^{196}Hg	^{197}Hg	^{198}Hg	^{199}Hg
atomic mass	195.97		197.97	198.97
natural abundance	0.2%	0%	10.1%	17.0%
half-life	stable	65 h	stable	stable

nuclide	^{200}Hg	^{201}Hg	^{202}Hg	^{204}Hg
atomic mass	199.97	200.97	201.97	203.97
natural abundance	23.1%	13.2%	29.6%	6.8%
half-life	stable	stable	stable	stable

Ionisation Energies/kJ mol^{-1}

M - M$^+$	1007
M$^+$ - M^{2+}	1809
M^{2+} - M^{3+}	3300
M^{3+} - M^{4+}	4400
M^{4+} - M^{5+}	5900
M^{5+} - M^{6+}	7400
M^{6+} - M^{7+}	9100
M^{7+} - M^{8+}	11600
M^{8+} - M^{9+}	13400
M^{9+} - M^{10+}	15300

Other Information

Enthalpy of Fusion/kJ mol^{-1} 2.33
Enthalpy of Vaporisation/kJ mol^{-1} 59.1

Oxidation States

main	Hg^{II}
other	Hg^{I}

Covalent Bonds /kJ mol^{-1}

not applicable

Molybdenum   *Mo*

General Information

Discovery

Molybdenum was discovered by P.J. Hjelm in 1781 in Uppsala, Sweden.

Appearance

The metal is silver-white and fairly soft when pure. It is usually obtained as a grey powder.

Source

The main source of this element is the ore molybdenite. Molybdenum can be obtained form this ore, but most commercial production is as a by-product of copper production.

Uses

Molybdenum is a valuable alloying agent, as it contributes to the hardness and toughness of quenched and tempered steels. It is also used in certain nickel-based alloys which are heat-resistant and corrosion-resistant to chemical solutions. It has found use in electrical and nuclear applications, and as a catalyst in the refining of petroleum.

Biological Role

Molybdenum is an essential element for animals and plants. If soil lacks this element the land is barren.

Physical Information

Atomic Number	42
Relative Atomic Mass ($^{12}C=12.000$)	95.94
Melting Point/K	2890
Boiling Point/K	4885
Density/kg m^{-3}	10220 (293K)
Ground State Electron Configuration	$[Kr]4d^5 5s^1$
Electron Affinity(M-M$^-$)/kJ mol^{-1}	114

Key Isotopes

nuclide	^{92}Mo	^{94}Mo	^{95}Mo	^{96}Mo
atomic mass	91.91	93.90	94.91	95.90
natural abundance	14.84%	9.25%	15.92%	16.68%
half-life	stable	stable	stable	stable

nuclide	^{97}Mo	^{98}Mo	^{99}Mo	^{100}Mo
atomic mass	96.91	97.91		99.91
natural abundance	9.55%	24.13%	0%	9.63%
half-life	stable	stable	66.69 hrs	stable

Ionisation Energies/kJ mol^{-1}

M - M$^+$	685
M$^+$ - M^{2+}	1558
M^{2+} - M^{3+}	2621
M^{3+} - M^{4+}	4480
M^{4+} - M^{5+}	5900
M^{5+} - M^{6+}	6560
M^{6+} - M^{7+}	12230
M^{7+} - M^{8+}	14800
M^{8+} - M^{9+}	16800
M^{9+} - M^{10+}	19700

Other Information

Enthalpy of Fusion/kJ mol^{-1} 27.6
Enthalpy of Vaporisation/kJ mol^{-1} 589.9

Oxidation States

main Mo^{VI}
others Mo^{-II}, Mo^{O}, Mo^{I}, Mo^{II}, Mo^{III}, Mo^{IV}, Mo^{V}

Covalent Bonds /kJ mol^{-1}

not applicable

Neodymium   *Nd*

General Information

Discovery

Neodymium was separated from the rare earth didymia by Baron Auer von Welsbach in 1885 in Vienna, Austria. The other principal component of didymia was praseodymium, atomic number 59.

Appearance

Neodymium is a bright silvery-white metal.

Source

The principal sources of most rare earth elements are the minerals monazite and bastnaesite. From these neodymium can extracted by ion exchange and solvent extraction techniques. The element can also be obtained by reducing the anhydrous chloride with calcium.

Uses

Neodymium is present in misch metal up to 18%. This alloy is used in such products as cigarette lighters where a light flint operates. Neodymium is also a component, along with praseodymium, of didymia, a special glass used in goggles in glass-blowing and welding. The element colours glass delicate shades of violet, wine-red and grey. It is used to make glass which transmits the tanning rays of the sun but not the harmful infrared rays.

Biological Role

Neodymium has no known biological role, is moderately toxic and a known eye irritant.

General Information

Neodymium reacts slowly with cold water and quickly with hot water. It quickly tarnishes in air and so is usually kept under paraffin or sealed in plastic.

It exists in two allotropic forms, with a transformation from hexagonal to body-centred cubic taking place at 863K.

Physical Information

Atomic Number	60
Relative Atomic Mass ($^{12}C=12.000$)	144.24
Melting Point/K	1294
Boiling Point/K	3341
Density/kg m^{-3}	7007 (293K)
Ground State Electron Configuration	[Xe]4f^{4}6s^2
Electron Affinity(M-M$^-$)/kJ mol^{-1}	50

Key Isotopes

nuclide	^{142}Nd	^{143}Nd	^{144}Nd	^{145}Nd
atomic mass	141.9	142.91	143.9	144.9
natural abundance	27.16%	12.18%	23.80%	8.29%
half-life	stable	stable	stable	stable

nuclide	^{146}Nd	^{147}Nd	^{148}Nd	^{150}Nd
atomic mass	145.9		147.9	149.9
natural abundance	17.19%	0%	5.75%	5.63%
half-life	stable	11 days	stable	stable

Ionisation Energies/kJ mol^{-1}

M - M$^+$	529.6
M$^+$ - M^{2+}	1035
M^{2+} - M^{3+}	2130
M^{3+} - M^{4+}	3899
M^{4+} - M^{5+}	
M^{5+} - M^{6+}	
M^{6+} - M^{7+}	
M^{7+} - M^{8+}	
M^{8+} - M^{9+}	
M^{9+} - M^{10+}	

Other Information

Enthalpy of Fusion/kJ mol^{-1}	7.11
Enthalpy of Vaporisation/kJ mol^{-1}	328

Oxidation States

main	Nd^{III}
others	Nd^{II}, Nd^{IV}

Covalent Bonds /kJ mol^{-1}

not applicable

Neon   Ne

General Information

Discovery

Neon was discovered by Sir William Ramsay and M.W. Travers in London in 1898.

Appearance

Neon is a colourless, odourless gas.

Source

Neon is a rare gas present in the atmosphere to the extent of 1 part in 65,000 of air. It is obtained by liquefaction of air and separation from other elements by fractional distillation.

Uses

In a vacuum discharge tube neon glows a reddish orange colour, and is therefore used in making the ubiquitous neon advertising signs, which accounts for its largest use. It is also used to make high-voltage indicators, lightning arrestors, wave meter tubes and television tubes. Liquid neon is now commercially available and is finding important application as an economic cryogenic refrigerant, as it has over 40 times more refrigerating capacity per unit volume than liquid helium and more than 3 times that of liquid hydrogen.

Biological Role

Neon has no known biological role. It is non-toxic.

General Information

Natural neon is a mixture of three isotopes, but five other unstable isotopes are known. It is a very inert element. Neon is said to form a compound with fluorine but it is still questionable whether such a compound truly exists.

Physical Information

Atomic Number	10
Relative Atomic Mass ($^{12}C=12.000$)	20.180
Melting Point/K	24.48
Boiling Point/K	27.10
Density/kg m^{-3}	0.900 (gas, 273K)
Ground State Electron Configuration	$[He]2s^2 2p^6$
Electron Affinity(M-M$^-$)/kJ mol^{-1}	-99

Key Isotopes

nuclide	^{20}Ne	^{21}Ne	^{22}Ne
atomic mass	19.992	20.993	21.991
natural abundance	90.51%	0.27%	9.22%
half-life	stable	stable	stable

Ionisation Energies/kJ mol^{-1}

M − M$^+$	2080.6
M$^+$ − M^{2+}	3952.2
M^{2+} − M^{3+}	6122
M^{3+} − M^{4+}	9370
M^{4+} − M^{5+}	12177
M^{5+} − M^{6+}	15238
M^{6+} − M^{7+}	19998
M^{7+} − M^{8+}	23069
M^{8+} − M^{9+}	115377
M^{9+} − M^{10+}	131429

Other Information

Enthalpy of Fusion/kJ mol^{-1} 0.324
Enthalpy of Vaporisation/kJ mol^{-1} 1.736

Oxidation States

not applicable

Covalent Bonds /kJ mol^{-1}

not applicable

Neptunium   *Np*

General Information

Discovery
Neptunium was discovered by E.M. McMillan and P. Abelson in 1940 in California, USA.

Appearance
Neptunium is a radioactive silvery metal.

Source
Neptunium is obtained as a by-product from nuclear reactors. Trace quantities occur naturally in uranium ores.

Uses
Neptunium is little used outside research.

Biological Role
Neptunium has no known biological role. It is toxic due to its radioactivity.

General Information
Neptunium is attacked by oxygen, steam and acids, but not by alkalis.

Physical Information

Atomic Number	93
Relative Atomic Mass ($^{12}C=12.000$)	237.05
Melting Point/K	913
Boiling Point/K	4175
Density/kg m^{-3}	20250 (293K)
Ground State Electron Configuration	[Rn]$5f^4 6d^1 7s^2$

Key Isotopes

nuclide ^{237}Np

atomic mass 237.05

natural abundance 0%

half-life 2.14×10^6 yrs

Ionisation Energies/kJ mol^{-1}

M   - M$^+$ 597

M$^+$ - M^{2+}

M^{2+} - M^{3+}

M^{3+} - M^{4+}

M^{4+} - M^{5+}

M^{5+} - M^{6+}

M^{6+} - M^{7+}

M^{7+} - M^{8+}

M^{8+} - M^{9+}

M^{9+} - M^{10+}

Other Information

Enthalpy of Fusion/kJ mol^{-1} 9.46

Enthalpy of Vaporisation/kJ mol^{-1} 336.6

Oxidation States

main   Np^V

others   Np^{II}, Np^{III}, Np^{IV}, Np^{VI}, Np^{VII}

Covalent Bonds /kJ mol^{-1}

not applicable

Nickel   *Ni*

General Information

Discovery

Nickel was discovered by A.F. Cronstedt in 1751 in Stockholm, Sweden.

Appearance

Nickel is a silvery-white metal which is lustrous, malleable and ductile. It is capable of taking on a high polish.

Source

The minerals which contain the most nickel are garnierite and pentlandite. About 30% of these minerals are found in Ontario in North America.

Uses

Nickel is chiefly used in the making of alloys such as stainless steel. A copper-nickel alloy is extensively used in making desalination plants for converting sea water into fresh water.

Nickel steel is used for armour plate.

Nickel has long been used in coins - the US five-cent piece is 25% nickel and 75% copper. Nickel plate protects softer metals. Finely-divided nickel is used as a catalyst for hydrogenating vegetable oils, and nickel imparts a green colour to glass.

Biological Role

The biological role of nickel is uncertain, but both the metal and nickel sulphide are considered to be carcinogenic. Nickel carbonyl is very toxic.

General Information

Nickel is very resistant to corrosion. It is soluble in all acids except concentrated nitric acid, and is not affected by alkalis. It is a fair conductor of heat and electricity.

Physical Information

Atomic Number	28
Relative Atomic Mass ($^{12}C=12.000$)	58.69
Melting Point/K	1726
Boiling Point/K	3005
Density/kg m^{-3}	8902 (298K)
Ground State Electron Configuration	$[Ar]3d^84s^2$
Electron Affinity(M-M$^-$)/kJ mol^{-1}	156

Key Isotopes

nuclide	^{58}Ni	^{59}Ni	^{60}Ni	^{61}Ni
atomic mass	57.935	58.934	59.933	60.931
natural abundance	68.27%	0%	26.10%	1.13%
half-life	stable	8×10^4yrs	stable	stable

nuclide	^{62}ni	^{63}Ni	^{64}Ni
atomic mass	61.928	62.930	63.928
natural abundance	3.59%	0%	0.91%
half-life	stable	92 yrs	stable

Ionisation Energies/kJ mol^{-1}

M - M$^+$	736.7
M$^+$ - M^{2+}	1753.0
M^{2+} - M^{3+}	3393
M^{3+} - M^{4+}	5300
M^{4+} - M^{5+}	7280
M^{5+} - M^{6+}	10400
M^{6+} - M^{7+}	12800
M^{7+} - M^{8+}	15600
M^{8+} - M^{9+}	18600
M^{9+} - M^{10+}	21660

Other Information

Enthalpy of Fusion/kJ mol^{-1}	17.6
Enthalpy of Vaporisation/kJ mol^{-1}	374.8

Oxidation States

main: Ni^{II}

others: Ni^{-I}, Ni^{O}, Ni^{I}, Ni^{III}, Ni^{IV}, Ni^{VI}

Covalent Bonds /kJ mol^{-1}

not applicable

Niobium Nb

General Information

Discovery

Niobium was discovered by C. Hatchett in 1801 in London, in an ore sent to England more than a century before by J. Winthrop, first Governor of Connecticut.

Appearance

Niobium is shiny, white, soft and ductile, and takes on a bluish sheen when exposed to air for a long time.

Source

The main source of this element is in the mineral columbite, which can be found in Canada, Brazil, the former USSR, Nigeria and elsewhere. However, it is commercially prepared as a by-product of tin extraction.

Uses

Niobium is used as an alloying agent in carbon and alloy steels and in non-ferrous metals, as it improves the strength of the alloy. It is also used in jet engines and rockets. This element has superconductive properties and is used in superconductive magnets which retain their properties in strong magnetic fields. This type of application could be used for the large-scale generation of electricity.

Biological Role

Niobium has no known biological role.

General Information

The name niobium was adopted officially in 1950 after years of controversy. The alternative name was columbium, and some metallugists still use this name.

Niobium resists corrosion due to an oxide film. It can be attacked by hot, concentrated acids but resists attack by fused alkalis. It starts to oxidise in air at 200K, and when processed at even moderate temperatures must be placed in a protective atmosphere.

Physical Information

Atomic Number	41
Relative Atomic Mass ($^{12}C=12.000$)	92.906
Melting Point/K	2741
Boiling Point/K	5015
Density/kg m^{-3}	8570 (293K)
Ground State Electron Configuration	[Kr]4d^{4}5s^1
Electron Affinity(M-M$^-$)/kJ mol^{-1}	109

Key Isotopes

nuclide	^{93}Nb	^{94}Nb
atomic mass	92.91	93.91
natural abundance	100%	0%
half-life	stable	2x10^{4}yrs

Ionisation Energies/kJ mol^{-1}

M - M$^+$	664
M$^+$ - M^{2+}	1382
M^{2+} - M^{3+}	2416
M^{3+} - M^{4+}	3695
M^{4+} - M^{5+}	4877
M^{5+} - M^{6+}	9899
M^{6+} - M^{7+}	12100
M^{7+} - M^{8+}	
M^{8+} - M^{9+}	
M^{9+} - M^{10+}	

Other Information

Enthalpy of Fusion/kJ mol^{-1} 27.2
Enthalpy of Vaporisation/kJ mol^{-1} 680.19

Oxidation States

main	Nb^V
others	Nb^{-III}, Nb^{-I}, Nb^I, Nb^{II}, Nb^{III}, Nb^{IV}

Covalent Bonds /kJ mol^{-1}

not applicable

Nitrogen   *N*

General Information

Discovery

Nitrogen was discovered by Daniel Rutherford in 1772 in Edinburgh, Scotland, but Scheele, Cavendish, Priestley and others about the same time studied "burnt or dephlogisticated air", as air without oxygen was then called.

Appearance

Nitrogen is a colourless, odourless gas.

Source

Nitrogen makes up 78% of the air, by volume. From this inexhaustable source it can be obtained by liquefaction and fractional distillation.

Uses

The largest consumer of nitrogen in our society is the ammonia industry - the Haber Process - to manufacture fertilisers. Large amounts of gas are also used by the electronics industry, which uses the gas as a blanketing medium during production of such components as transistors, diodes etc. Large quantites of nitrogen are used in annealing stainless steel and other steel mill products. The drug industry also uses large quantities. Nitrogen is used as a refrigerant both for the immersion freezing of food products and for the transportation of food. Liquid nitrogen is also used in missile work and by the oil industry to build up great pressures in wells to force crude oil upwards.

Biological Role

Nitrogen is the basis of life as part of the DNA molecule.

General Information

The element nitrogen is so inert that Lavoisier named it 'azote', meaning 'without life', yet its compounds are so active as to be most important in many essential foods, fertilisers, poisons and explosives.

When nitrogen is heated , it combines directly with magnesium, lithium and calcium. When mixed with oxygen and subjected to electric sparks, it forms first nitrogen monoxide and then nitrogen dioxide. When mixed with hydrogen and heated under pressure, ammonia is formed (the Haber Process).

Physical Information

Atomic Number	7
Relative Atomic Mass ($^{12}C=12.000$)	14.007
Melting Point/K	63.29
Boiling Point/K	77.4
Density/kg m^{-3}	1.25 (gas, 273K)
Ground State Electron Configuration	[He]$2s^2 2p^3$
Electron Affinity(M-M$^-$)/kJ mol^{-1}	-31

Key Isotopes

nuclide	^{14}N	^{15}N
atomic mass	14.003	15.000
natural abundance	99.63%	0.37%
half-life	stable	stable

Ionisation Energies/kJ mol^{-1}

M - M$^+$	1402.3
M$^+$ - M^{2+}	2856.1
M^{2+} - M^{3+}	4578.0
M^{3+} - M^{4+}	7474.9
M^{4+} - M^{5+}	9440.0
M^{5+} - M^{6+}	53265.6
M^{6+} - M^{7+}	64358.7

Other Information

Enthalpy of Fusion/kJ mol^{-1}	0.720
Enthalpy of Vaporisation/kJ mol^{-1}	5.577

Oxidation States

N^{-III}, N^{-II}, N^{-I}, N^{O}, N^{II}, N^{III}, N^{IV}, N^{V}

Covalent Bonds /kJ mol^{-1}

N-H	390
N-N	160
N=N	415
N≡N	946
N-Cl	193

Nobelium   *No*

General Information

Discovery

The discovery of element 102 was disputed, but it was conclusively identified in 1958 by A. Ghiorso and co-workers in California, USA.

Appearance

Nobelium is a radioactive metal. Only a few atoms have ever been made, so its appearance and properties are unknown.

Source

Nobelium is made by the bombardment of curium with carbon nuclei.

Uses

Nobelium has no uses outside research.

Biological Role

Nobelium has no known biological role. It is toxic due to its radioactivity.

Physical Information

Atomic Number	102
Relative Atomic Mass ($^{12}C=12.000$)	259 (radioactive)
Melting Point/K	not available
Boiling Point/K	not available
Density/kg m^{-3}	not available
Ground State Electron Configuration	$[Rn]5f^{14}7s^2$
Electron Affinity(M-M$^-$)/kJ mol^{-1}	not available

Key Isotopes

nuclide ^{259}No

atomic mass

natural abundance 0%

half-life 58 mins

Ionisation Energies/kJ mol^{-1}

M   - M$^+$ 642
M$^+$ - M^{2+}
M^{2+} - M^{3+}
M^{3+} - M^{4+}
M^{4+} - M^{5+}
M^{5+} - M^{6+}
M^{6+} - M^{7+}
M^{7+} - M^{8+}
M^{8+} - M^{9+}
M^{9+} - M^{10+}

Other Information

Enthalpy of Fusion/kJ mol^{-1}

not available

Enthalpy of Vaporisation/kJ mol^{-1}

not available

Oxidation States

main   No^{II}
other   No^{III}

Covalent Bonds /kJ mol^{-1}

not applicable

Osmium Os

General Information

Discovery

Osmium was discovered by S. Tennant in 1803 in London.

Appearance

Osmium is lustrous, bluish-white, extremely hard and has a pungent smell.

Source

Osmium occurs in the free state and in the mineral osmiridium, but commercial recovery is from the wastes of nickel refining.

Uses

Osmium is almost entirely used to produce very hard alloys for fountain pen tips, instrument pivots, needles and electrical contacts.

Biological Role

Osmium has no known biological role, but is very toxic, and can cause lung, skin and eye damage.

General Information

Osmium metal is unaffected by air, water and acids, but dissolves in molten alkalis. The powdered metal slowly gives off osmium (VIII) oxide, the source of its pungent odour.

Physical Information

Atomic Number	76
Relative Atomic Mass ($^{12}C=12.000$)	190.2
Melting Point/K	3327
Boiling Point/K	5300
Density/kg m^{-3}	22590 (293K)
Ground State Electron Configuration	[Xe]$4f^{14}5d^{6}6s^{2}$
Electron Affinity(M-M$^-$)/kJ mol^{-1}	139

Key Isotopes

nuclide	^{184}Os	^{185}Os	^{186}Os	^{187}Os	^{188}Os
atomic mass	183.9		185.9	186.9	187.9
natural abundance	0.02%	0%	1.58%	1.6%	13.3%
half-life	stable	9.6 days	stable	stable	stable

nuclide	^{189}Os	^{190}Os	^{191}Os	^{192}Os
atomic mass	188.9	189.9		191.9
natural abundance	16.1%	26.4%	0%	41%
half-life	stable	stable	15 days	stable

Ionisation Energies/kJ mol^{-1}

M − M$^+$	840
M$^+$ − M^{2+}	1600
M^{2+} − M^{3+}	2400
M^{3+} − M^{4+}	3900
M^{4+} − M^{5+}	5200
M^{5+} − M^{6+}	6600
M^{6+} − M^{7+}	8100
M^{7+} − M^{8+}	9500
M^{8+} − M^{9+}	
M^{9+} − M^{10+}	

Other Information

Enthalpy of Fusion/kJ mol^{-1} 29.3
Enthalpy of Vaporisation/kJ mol^{-1} 738

Oxidation States
main   Os^{IV}
others   Os^{-II}, Os^{0}, Os^{I}, Os^{II}, Os^{III}, Os^{V}, Os^{VI}, Os^{VII}, Os^{VIII}

Covalent Bonds /kJ mol^{-1}
not applicable

Oxygen   O

General Information

Discovery

For many centuries, workers occasionally realised that air was composed of more than one component. The behaviour of oxygen and nitrogen as components of air led to the advancement of the phlogiston theory of combustion. Oxygen was prepared by several workers, including Bayen and Borch, but they did not recognise it as an element. Its discovery is generally credited to J. Priestley in 1774, in Leeds, and independently to C.W. Scheele in Uppsala, Sweden.

Appearance

Oxygen is a colourless, odourless, tasteless gas.

Source

Oxygen, as a gaseous element, forms 21% of the atmosphere by volume from which it can be obtained by liquefaction and fractional distillation. The element and its compounds make up 49.2%, by weight, of the earth's crust. About two thirds of the human body and nine tenths of water is oxygen. In the laboratory it can be prepared by the electrolysis of water or by heating potassium chlorate with manganese dioxide (manganese (IV) oxide) as a catalyst.

Uses

Oxygen is very reactive and capable of combining with most other elements. It is a component of thousands of organic compounds, and is essential for the respiration of all plants and animals and for almost all combustion.

The greatest commercial use of gaseous oxygen is oxygen enrichment of steel blast furnaces. Large quantities are also used in making synthesis gas for ammonia and methanol, ethylene oxide, and for oxy-acetylene welding.

Biological Role

Oxygen is the basis of all life as part of the DNA molecule. It is breathed in by animals and restored to the air by the photosynthesis mechanism of plants.

General Information

The liquid and solid forms of oxygen are pale blue in colour and strongly paramagnetic. Ozone is a highly active allotropic form of oxygen, and is formed by the action of an electrical discharge or ultraviolet light on oxygen. The presence of ozone in the atmosphere (amounting to the equivalent of a layer 3mm thick at ordinary temperatures and pressures) is of vital importance in preventing harmful ultraviolet rays of the sun from reaching the surface of the earth. Recently, concern has mounted that the use of aerosols is reducing the thickness of this ozone layer.

Physical Information

Atomic Number	8
Relative Atomic Mass ($^{12}C=12.000$)	15.999
Melting Point/K	54.8
Boiling Point/K	90.188
Density/kg m^{-3}	1.429 (gas, 273K)
Ground State Electron Configuration	[He]$2s^2\,2p^4$
Electron Affinity/kJ mol^{-1} O→O$^-$	141
O$^-$→O^{2-}	-703

Key Isotopes

nuclide	^{16}O	^{17}O	^{18}O
atomic mass	1994	16.999	17.999
natural abundance	99.76%	0.038%	0.200%
half-life	stable	stable	stable

Ionisation Energies/kJ mol^{-1}

M - M$^+$	1313.9
M$^+$ - M^{2+}	3388.2
M^{2+} - M^{3+}	5300.3
M^{3+} - M^{4+}	7469.1
M^{4+} - M^{5+}	10989.3
M^{5+} - M^{6+}	13326.2
M^{6+} - M^{7+}	71333.3
M^{7+} - M^{8+}	84076.3

Other Information

Enthalpy of Fusion/kJ mol^{-1} 0.444
Enthalpy of Vaporisation/kJ mol^{-1} 6.82

Oxidation States

main   O^{-II}
others O^{-I}, O^{0}, O^{I}, O^{II}

Covalent Bonds

O-O	146
O=O	498
N-O	200

Palladium — Pd

General Information

Discovery
Palladium was discovered by W.H. Wollaston in 1803 in London.

Appearance
Palladium is a steel-white metal which is lustrous, malleable and ductile. It does not tarnish in air.

Source
It is found associated with platinum and other metals in deposits in the former USSR, North and South America and Australia. It is also found associated with nickel-copper deposits in South Africa and USA. It is extracted commercially from these latter ores.

Uses
Finely divided palladium is a good catalyst and is used for hydrogenation and dehydrogenation reactions. White gold is an alloy of gold decolourised by the addition of palladium. It is also used with gold, silver and other metals as a "stiffener" in dental inlays and bridgework. Hydrogen easily diffuses through heated palladium and this provides a way of purifying the gas.

Biological Role
Palladium has no known biological role, and is non-toxic.

General Information
Palladium resists corrosion, but dissolves in oxidising acids and fused alkalis. At room temperature the metal has the unusual property of absorbing up to 900 times its own volume of hydrogen.

Physical Information

Atomic Number	46
Relative Atomic Mass ($^{12}C=12.000$)	106.42
Melting Point/K	1825
Boiling Point/K	3413
Density/kg m^{-3}	12020 (293K)
Ground State Electron Configuration	[Kr]4d^{10}
Electron Affinity(M-M$^-$)/kJ mol^{-1}	98.4

Key Isotopes

nuclide	^{102}Pd	^{103}Pd	^{104}Pd	^{105}Pd
atomic mass	101.901		103.90	104.90
natural abundance	1.02%	0%	11.14%	22.33%
half-life	stable	17 days	stable	stable

nuclide	^{106}Pd	^{108}Pd	^{109}Pd	^{110}Pd
atomic mass	105.90	107.90		
natural abundance	27.33%	26.46%	0%	11.72%
half-life	stable	stable	13.47 h	stable

Ionisation Energies/kJ mol^{-1}

M - M$^+$	805
M$^+$ - M^{2+}	1875
M^{2+} - M^{3+}	3177
M^{3+} - M^{4+}	4700
M^{4+} - M^{5+}	6300
M^{5+} - M^{6+}	8700
M^{6+} - M^{7+}	10700
M^{7+} - M^{8+}	12700
M^{8+} - M^{9+}	15000
M^{9+} - M^{10+}	17200

Other Information

Enthalpy of Fusion/kJ mol-1   17.2
Enthalpy of Vaporisation/kJ mol-1   361.5

Oxidation States

main   Pd^{II}
others   Pd^{O}, Pd^{IV}

Covalent Bonds /kJ mol^{-1}

not applicable

Phosphorus   P

General Information

Discovery

Phosphorus was discovered in 1669 by H. Brandt in Hamburg, Germany, by extraction from urine.

Appearance

Phosphorus occurs in three major forms, white, red and black. The white form appears as a waxy white solid, but when pure is colourless and transparent. The red and black forms are powders of the appropriate colour.

Source

Phosphorus is not found free in nature, but is widely distributed in combination with minerals. An important source is phosphate rock, which contains the apatite minerals and is found in large quantities in the USA, the "USSR" and elsewhere.

White phosphorus may be made commercially by several methods. Usually phosphate rock is heated in the presence of carbon and silica in a furnace, which produces phosphorus as a vapour which is then collected under water. It can then be converted to red phosphorus by heating for several days.

Uses

Fertilisers contain a high proportion of phosphorus and are manufactured from concentrated phosphoric acids. World wide demand for fertilisers has greatly increased in recent years as their importance to agriculture and farming has grown. Phosphorus is also important in the production of steel.

Phosphates are ingredients of some detergents, but are increasingly being omitted nowadays due to concern that high phosphate levels in natural water supplies cause the growth of undesirable algae. Phosphates are also used in the production of special glasses and fine chinaware.

Biological Role

Phosphorus is the basis of life as part of the DNA molecule. White phosphorus is very toxic and contact with skin can cause severe burns.

General Information

White phosphorus is insoluble in water but soluble in carbon disulphide. It burns spontaneously in air. When exposed to sunlight or heated in its own vapour to 250K it is converted to red phosphorus, which is less dangerous than the white form and does not ignite spontaneously. Red phosphorus is used in the manufacture of safety matches, pesticides, incendiary shells, smoke bombs and tracer pellets.

Physical Information

Atomic Number	15
Relative Atomic Mass ($^{12}C=12.000$)	30.974
Melting Point/K	317.3 (white), 683 (red)
Boiling Point/K	553 (white)
Density/kg m^{-3}	1820 (white)
	2200 (red)
	2690 (black), all at 293K
Ground State Electron Configuration	[Ne]$3s^2 3p^3$
Electron Affinity(M-M$^-$)/kJ mol^{-1}	60

Key Isotopes

nuclide	^{31}P	^{32}P	^{33}P
atomic mass	30.974	31.974	32.972
natural abundance	100%	0%	0%
half-life	stable	14.3 days	25 days

Ionisation Energies/kJ mol^{-1}

M - M$^+$	1011.7
M$^+$ - M^{2+}	1903.2
M^{2+} - M^{3+}	2912
M^{3+} - M^{4+}	4956
M^{4+} - M^{5+}	6273
M^{5+} - M^{6+}	21268
M^{6+} - M^{7+}	25397
M^{7+} - M^{8+}	29854
M^{8+} - M^{9+}	35867
M^{9+} - M^{10+}	40958

Other Information

Enthalpy of Fusion/kJ mol^{-1}
2.51 (white)

Enthalpy of Vaporisation/kJ mol^{-1}
51.9 (white)

Oxidation States

main	P^V
others	P^{-III}, P^{-II}, P^O, P^{II}, P^{III}

Covalent Bonds /kJ mol^{-1}

P-H	328
P-O	407
P=O	560
P-F	490
P-Cl	319
P-P	209

Platinum   *Pt*

General Information

Discovery

Platinum was discovered by South Americans and taken to Europe about 1750. The metal was used by pre-Columbian Indians.

Appearance

Platinum is a beautiful silvery-white metal, and is malleable and ductile.

Source

Platinum is found uncombined in alluvial deposits, and prepared commercially as a by-product of nickel refining from copper-nickel ores.

Uses

Platinum is used extensively for jewellery, wire and many valuable instruments including thermocouple elements. It is also used for electrical contacts, corrosion-resistance apparatus and in dentistry. In a finely divided state platinum absorbs large volumes of hydrogen and so is used as a catalyst in the petroleum cracking industry.

Biological Role

Platinum has no known biological role, and is non-toxic.

General Information

Platinum is not affected by air or water at any temperature. It is insoluble in hydrochloric and nitric acids, but dissolves when they are mixed as aqua regia.

The price of platinum fluctuates, but it can cost eight times as much as gold.

Physical Information

Atomic Number	78
Relative Atomic Mass ($^{12}C=12.000$)	195.08
Melting Point/K	2045
Boiling Point/K	4100
Density/kg m^{-3}	21450 (293K)
Ground State Electron Configuration	[Xe]$4f^{14}5d^{9}6s^{1}$
Electron Affinity(M-M$^-$)/kJ mol^{-1}	247

Key Isotopes

nuclide	^{190}Pt	^{192}Pt	^{194}Pt	^{195}Pt
atomic mass	189.96	191.96	193.96	194.96
natural abundance	0.01%	0.79%	32.9%	33.8%
half-life	6.9x10^{11}yrs	10^{15} yrs	stable	stable

nuclide	^{196}Pt	^{197}Pt	^{198}Pt
atomic mass	195.96		197.97
natural abundance	25.3%	0%	7.2%
half-life	stable	18 h	stable

Ionisation Energies/kJ mol^{-1}

M - M$^+$	870
M$^+$ - M^{2+}	1791
M^{2+} - M^{3+}	2800
M^{3+} - M^{4+}	3900
M^{4+} - M^{5+}	5300
M^{5+} - M^{6+}	7200
M^{6+} - M^{7+}	8900
M^{7+} - M^{8+}	10500
M^{8+} - M^{9+}	12300
M^{9+} - M^{10+}	14100

Other Information

Enthalpy of Fusion/kJ mol^{-1} 19.7
Enthalpy of Vaporisation/kJ mol^{-1} 469

Oxidation States

main   Pt^{IV}
others   Pt^{O}, Pt^{II}, Pt^{V}, Pt^{VI}

Covalent Bonds /kJ mol^{-1}

not applicable

Plutonium *Pu*

General Information

Discovery

Plutonium was discovered by G.T. Seaborg, A.C. Wahl and J. W. Kennedy in 1940 in California, USA.

Appearance

Plutonium is a radioactive silvery metal that tarnishes in air to give an oxide coating with yellow tinge.

Source

The greatest source of plutonium - and one that produces 20,000 kilograms every year - is the irradiation of uranium in nuclear reactors. This produces the isotope ^{239}Pu, with a half-life of 24400 years.

Uses

Plutonium was used in several of the first atomic bombs, and is still used for explosives. The complete detonation of a kilogram of plutonium produces an explosion equivalent to over 10000 tonnes of chemical explosive. Plutonium is also a key material in the development of nuclear power. It has been used as a compact energy source on space missions such as the Apollo lunar missions.

Biological Role

Plutonium has no known biological role. It is toxic due to its radioactivity.

General Information

Plutonium is attacked by oxygen, steam and acids, but not by alkalis. The metal is warm to the touch because of the energy given off in alpha decay, and a large piece of the metal can boil water. Plutonium forms compounds with oxygen, the halides, carbon, nitrogen and silicon.

Physical Information

Atomic Number	94
Relative Atomic Mass ($^{12}C=12.000$)	244 (radioactive)
Melting Point/K	914
Boiling Point/K	3505
Density/kg m^{-3}	19840 (298K)
Ground State Electron Configuration	[Rn]$5f^6 7s^2$

Key Isotopes

nuclide	^{239}Pu	^{242}Pu	^{244}Pu
atomic mass	239.05	242.06	244.06
natural abundance	0%	0%	0%
half-life	24400 yrs	3.79×10^5 yrs	8.2×10^7 yrs

Ionisation Energies/kJ mol^{-1}

M − M$^+$	585
M$^+$ − M^{2+}	
M^{2+} − M^{3+}	
M^{3+} − M^{4+}	
M^{4+} − M^{5+}	
M^{5+} − M^{6+}	
M^{6+} − M^{7+}	
M^{7+} − M^{8+}	
M^{8+} − M^{9+}	
M^{9+} − M^{10+}	

Other Information

Enthalpy of Fusion/kJ mol^{-1}	2.8
Enthalpy of Vaporisation/kJ mol^{-1}	343.5

Oxidation States

main	Pu^{IV}
others	Pu^{II}, Pu^{III}, Pu^V, Pu^{VI}, Pu^{VII}

Covalent Bonds /kJ mol^{-1}

not applicable

Polonium   Po

General Information

Discovery

Polonium was discovered in 1898 in Paris, France. It was the first element discovered by Marie Curie, while she was investigating the cause of radioactivity in pitchblende.

Appearance

Polonium is a silvery-grey, radioactive metal.

Source

Polonium is a very rare natural element. It is obtained when natural bismuth, ^{209}Bi, is bombarded by neutrons to give ^{210}Bi, the parent of polonium.

Uses

Polonium is an alpha-emitter, and is used as an alpha-particle source for scientific research in the form of a thin film on a stainless steel disc. It is also used as a heat source in space equipment. It can be mixed or alloyed with beryllium to provide a source of neutrons.

Biological Role

Polonium has no known biological role. It is highly toxic due to its radioactivity.

General Information

Polonium is readily dissolved in dilute acids, but is only slightly soluble in alkalis. A milligram of polonium emits as many alpha particles as 5 grams of radium. The energy released by its decay is so large that a capsule containing about 0.5 grams reaches a temperature above 500K.

Physical Information

Atomic Number	84
Relative Atomic Mass (^{12}C=12.000)	209 (radioactive)
Melting Point/K	527
Boiling Point/K	1235
Density/kg m^{-3}	9320 (293K)
Ground State Electron Configuration	[Xe]4f^{14}5d^{10}6s^{2}6p^4
Electron Affinity(M-M$^-$)/kJ mol^{-1}	186

Key Isotopes

nuclide	^{209}Po	^{210}Po	^{211}Po	^{216}Po	^{218}Po
atomic mass	208.98	209.98	210.99	216.0	218.0
natural abundance	0%	trace	trace	trace	trace
half-life	103 yrs	138.4 days	0.52 secs	0.15 secs	3.05 mins

Ionisation Energies/kJ mol^{-1}

M – M$^+$	812
M$^+$ – M^{2+}	1800
M^{2+} – M^{3+}	2700
M^{3+} – M^{4+}	3700
M^{4+} – M^{5+}	5900
M^{5+} – M^{6+}	7000
M^{6+} – M^{7+}	10800
M^{7+} – M^{8+}	12700
M^{8+} – M^{9+}	14900
M^{9+} – M^{10+}	17000

Other Information

Enthalpy of Fusion/kJ mol^{-1} 10
Enthalpy of Vaporisation/kJ mol^{-1} 100.8

Oxidation States

main   Po^{IV}
others   Po^{-II}, Po^{II}, Po^{VI}

Covalent Bonds /kJ mol^{-1}

not applicable

Potassium   K

General Information

Discovery

Potassium was discovered by Sir Humphry Davy in 1807 in London, by the electrolysis of potassium hydroxide (potash). This was the first metal to be isolated by electrolysis.

Appearance

Potassium is a soft, white metal which is silvery when cut but which rapidly oxidises.

Source

Potassium is the seventh most abundant metal and makes up 2.4% by mass of the earth's crust. Most minerals containing potassium are sparingly soluble and the metal is difficult to obtain from them. Certain minerals, however, such as sylvite, sylvinite and carnallite are found in ancient water beds and potassium salts can be easily recovered from these vast deposits. Potassium hydroxide, potash, is mined in several places including Germany and the USA. Potassium is also found in the ocean in small amounts compared to sodium.

Uses

The greatest demand for potassium is the use of potash in fertilisers. Many other potassium salts are of great importance, including the nitrate, carbonate, chloride, bromide, cyanide and sulphate.

Biological Role

Potassium is essential to life, and non-toxic. One of its natural isotopes is radioactive, and although this radioactivity is mild, it may be one natural cause of genetic mutation in man.

General Information

Potassium is the lightest known metal. It is also one of the most reactive and electropositive of metals, and as it oxidises rapidly in air it must be preserved in a mineral oil such as kerosene. Its reaction with water is vigorous - it catches fire spontaneously and decomposes with the evolution of hydrogen. Potassium and its salts burn with a violet colour.

Physical Information

Atomic Number	19
Relative Atomic Mass ($^{12}C=12.000$)	39.098
Melting Point/K	336.80
Boiling Point/K	1047
Density/kg m^{-3}	862 (293K)
Ground State Electron Configuration	[Ar]4s^1
Electron Affinity(M-M$^-$)/kJ mol^{-1}	43.8

Key Isotopes

nuclide	^{39}K	^{40}K	^{41}K	^{42}K	^{43}K
atomic mass	38.964	39.974	40.962	41.963	42.964
natural abundance	93.258%	0.0117%	6.730%	0%	0%
half-life	stable	1.28x10^{9}yrs	stable	12 h	22.4hrs

Ionisation Energies/kJ mol^{-1}

M - M$^+$	418.8
M$^+$ - M^{2+}	3051.4
M^{2+} - M^{3+}	4411
M^{3+} - M^{4+}	5877
M^{4+} - M^{5+}	7975
M^{5+} - M^{6+}	9649
M^{6+} - M^{7+}	11343
M^{7+} - M^{8+}	14942
M^{8+} - M^{9+}	16964
M^{9+} - M^{10+}	48575

Other Information

Enthalpy of Fusion/kJ mol^{-1}	2.40
Enthalpy of Vaporisation/kJ mol^{-1}	79.1

Oxidation States

main	K^I
other	K^{-I}

Covalent Bonds /kJ mol^{-1}

not applicable

Praseodymium *Pr*

General Information

Discovery

Praseodymium was first separated from the rare earth didymia by Baron Auer von Welsbach in 1885 in Vienna, Austria. The other component of didymia was neodymium, atomic number 60.

Appearance

Praseodymium is a soft, malleable, silvery metal.

Source

Praseodymium occurs along with other rare-earth elements in a variety of minerals, but the two principal commercial sources of most of these elements are monazite and bastnaesite. The usual techniques employed are ion exchange and solvent extraction, although praseodymium is also prepared by calcium reduction of the anhydrous chloride.

Uses

Praseodymium comprises 5% of the alloy misch metal, which is used in making products such as cigarette lighters. Along with other rare earth elements it is used in carbon arcs for studio lighting and projection. Praseodymium is also a component of didymium glass, used by welders and glassmakers, because it filters out the yellow light present in glass blowing. Salts of this element are used to colour glasses and enamels an intense and unusually clean yellow.

Biological Role

Praseodymium has no known biological role, and low toxicity.

General Information

Praseodymium reacts rapidly with water and slowly with oxygen to give a green oxide coating. It is stored under paraffin or sealed in plastic.

Physical information

Atomic Number	59
Relative Atomic Mass ($^{12}C=12.000$)	140.91
Melting Point/K	1204
Boiling Point/K	3785
Density/kg m^{-3}	6773 (293K)
Ground State Electron Configuration	[Xe]4f^{3}6s^2
Electron Affinity(M-M$^-$)/kJ mol^{-1}	50

Key Isotopes

nuclide	^{141}Pr	^{142}Pr	^{143}Pr
atomic mass	140.91		
natural abundance	100%	0%	0%
half-life	stable	19.2 h	13.59 days

Ionisation Energies/kJ mol^{-1}

M - M$^+$	523.1
M$^+$ - M^{2+}	1018
M^{2+} - M^{3+}	2086
M^{3+} - M^{4+}	3761
M^{4+} - M^{5+}	5543
M^{5+} - M^{6+}	
M^{6+} - M^{7+}	
M^{7+} - M^{8+}	
M^{8+} - M^{9+}	
M^{9+} - M^{10+}	

Other Information

Enthalpy of Fusion/kJ mol^{-1}	11.3
Enthalpy of Vaporisation/kJ mol^{-1}	357

Oxidation States

main	Pr^{III}
other	Pr^{IV}

Covalent Bonds /kJ mol^{-1}

not applicable

Promethium *Pm*

General Information

Discovery

The existence of promethium was predicted by Branner in 1902. In 1945 the element was first produced by the irradiation of neodymium by J.A. Marinsky, L.E. Glendenin and C.D. Coryell in Oak Ridge, USA.

Appearance

Promethium is a radioactive metal. Its salts luminesce in the dark with a pale greenish glow.

Source

Promethium is not found on the planet Earth. It has been identified on Andromeda. It can be produced by the irradiation of neodymium and praseodymium with neutrons, deuterons and alpha particles. It can also be prepared by ion exchange of atomic reactor fuel processing wastes.

Uses

Promethium is used as a nuclear-powered battery as it can capture light in photocells and convert it into an electric current. Such batteries are used in watches, radios and guided-missile instruments. They are no larger than a drawing pin.

Biological Role

Promethium has no known biological role, but is toxic due to its radioactivity.

General Information

Little is known about the properties of promethium.

Physical Information

Atomic Number	61
Relative Atomic Mass ($^{12}C=12.000$)	145 (radioactive)
Melting Point/K	1441
Boiling Point/K	ca. 3000
Density/kg m^{-3}	7220 (298K)
Ground State Electron Configuration	[Xe]4f^{5}6s^2
Electron Affinity(M-M$^-$)/kJ mol^{-1}	50

Key Isotopes

nuclide	^{145}Pm	^{146}Pm	^{147}Pm	^{149}Pm	^{151}Pm
atomic mass	144.9		146.9		
natural abundance	0%	0%	0%	0%	0%
half-life	17.7 yrs	4.4 yrs	2.62 yrs	53.1 h	28 h

Ionisation Energies/kJ mol^{-1}

M − M$^+$	535.9
M$^+$ − M^{2+}	1052
M^{2+} − M^{3+}	2150
M^{3+} − M^{4+}	3970
M^{4+} − M^{5+}	
M^{5+} − M^{6+}	
M^{6+} − M^{7+}	
M^{7+} − M^{8+}	
M^{8+} − M^{9+}	
M^{9+} − M^{10+}	

Other Information

Enthalpy of Fusion/kJ mol^{-1} 12.6

Enthalpy of Vaporisation/kJ mol^{-1}

not applicable

Oxidation State   Pm^{III}

Covalent Bonds /kJ mol^{-1}

not applicable

Protactinium   Pa

General Information

Discovery

Protactinium was discovered in 1917 by Hahn and Meitner in Berlin, Fajans in Germany and Fleck in Glasgow. It was initially named brevium, as the first isotope identified was very short-lived.

Appearance

Protactinium is a radioactive, silvery metal.

Source

Protactinium is found naturally in uranium ores and produced in gram quantities from uranium fuel elements.

Uses

Protactinium is little used.

Biological Role

Protactinium has no known biological role. It is toxic due to its radioactivity.

General Information

Protactinium is attacked by oxygen, steam and acids, but not by alkalis. It is the third rarest of the elements.

Physical Information

Atomic Number	91
Relative Atomic Mass ($^{12}C=12.000$)	231.04
Melting Point/K	2113
Boiling Point/K	4300
Ground State Electron Configuration	$[Rn]5f^26d^17s^2$

Key Isotopes

nuclide	^{231}Pa	^{233}Pa	^{234}Pa
atomic mass	231.04	233.04	234.04
natural abundance	trace	0%	trace
half-life	3.26×10^4 yrs	27 days	6.75 h

Ionisation Energies/kJ mol^{-1}

M - M$^+$	568
M$^+$ - M^{2+}	
M^{2+} - M^{3+}	
M^{3+} - M^{4+}	
M^{4+} - M^{5+}	
M^{5+} - M^{6+}	
M^{6+} - M^{7+}	
M^{7+} - M^{8+}	
M^{8+} - M^{9+}	
M^{9+} - M^{10+}	

Other Information

Enthalpy of Fusion/kJ mol^{-1} 16.7
Enthalpy of Vaporisation/kJ mol^{-1} 481

Oxidation States

main   Pa^V
others   Pa^{III}, Pa^{IV}

Covalent Bonds /kJ mol^{-1}

not applicable

Radium   *Ra*

General Information

Discovery

Radium was discovered by Pierre and Marie Curie in 1898 in Paris, France, from pitchblende. There is about 1 gram of radium in 7 tonnes of pitchblende.

It was isolated in 1911 by Marie Curie and Debierne, by the electrolysis of a solution of pure radium chloride.

Appearance

Pure radium is brilliant white when freshly prepared, but blackens on exposure to the air. The metal and its salts luminesce.

Source

Radium is present in all uranium ores, and could be extracted as a by-product of uranium refining. The usual source of pitchblende comes from Bohemia, but some radium-containing ores are found in Canada and the USA. Annual production of this element is less than 100 grams.

Uses

Radium was formerly used in the production of luminous paints, but this is now considered too hazardous. The element gives off small amounts of radium gas which has been used to treat cancer, but this use is now also considered too toxic - other radioactive sources are more powerful and safer to use.

Biological Role

Radium has no known biological role. It is toxic due to its radioactivity.

General Information

Radium reacts with both oxygen and water, and is somewhat more volatile than barium. It burns with a carmine red colour.

Radium emits alpha, beta and gamma rays. The final product of its disintigration is lead.

Physical Information

Atomic Number	88
Relative Atomic Mass ($^{12}C=12.000$)	226.02
Melting Point/K	973
Boiling Point/K	1413
Density/kg m^{-3}	5000 (293K)
Ground State Electron Configuration	[Rn]7s^2

Key Isotopes

nuclide	^{223}Ra	^{224}Ra	^{226}Ra	^{228}Ra
atomic mass	223.02	224.02	226.03	228.03
natural abundance	some	some	some	some
half-life	11.43 days	3.64 days	1602 yrs	5.77 yrs

Ionisation Energies/kJ mol^{-1}

M - M$^+$	509.3
M$^+$ - M^{2+}	979
M^{2+} - M^{3+}	3300
M^{3+} - M^{4+}	4400
M^{4+} - M^{5+}	5700
M^{5+} - M^{6+}	7300
M^{6+} - M^{7+}	8600
M^{7+} - M^{8+}	9900
M^{8+} - M^{9+}	13500
M^{9+} - M^{10+}	15100

Other Information

Enthalpy of Fusion/kJ mol^{-1}	7.15
Enthalpy of Vaporisation/kJ mol^{-1}	136.7

Oxidation State Ra^{II}

Covalent Bonds /kJ mol^{-1}

not applicable

Radon   *Rn*

General Information

Discovery

Radon was discovered by F.E. Dorn in 1900 in Halle, Germany, who named it radium emantium. The element was isolated in 1908 by Ramsay and Gray, who named it niton. Since 1923 it has been called radon.

Appearance

Radon is a colourless, odourless inert gas.

Source

Radon is produced naturally from the decay of a radium isotope, ^{226}Ra.

Uses

Radon decays into radioactive polonium and alpha rays, and this emitted radiation makes radon useful in cancer therapy. The gas is sealed in minute tubes called seeds or needles and implanted into the site of a tumour. The diseased tissue is thus destroyed in situ by the radiation.

Biological Role

Radon has no known biological role. It is toxic due to its radioactivity, the main hazard arising from inhalation, as the element and its radioactive daughters collect on dust particles.

General Information

Radon is the densest known gas.

Chemically, radon should resemble xenon but it has been little studied because any compounds which are formed are destroyed by hazardous radiation. It is reported that radon reacts with fluorine to give radon fluoride, and radon clathrates have also been reported. At ordinary temperatures radon is a colourless gas, but when cooled below freezing point it exhibits a brilliant phosphorescence which becomes yellow as the temperature is lowered and orange at the temperature of liquid air.

Physical Information

Atomic Number	86
Relative Atomic Mass ($^{12}C=12.000$)	222 (radioactive)
Melting Point/K	202
Boiling Point/K	211.4
Density/kg m^{-3}	9.73 (gas, 273K)
Ground State Electron Configuration	[Xe]4f^{14}5d^{10}6s^{2}6p^6
Electron Affinity(M-M$^-$)/kJ mol^{-1}	-41

Key Isotopes

nuclide	^{219}Rn	^{220}Rn	^{222}Rn
atomic mass	219.01	220.01	222.02
natural abundance	trace	trace	trace
half-life	4 secs	55 secs	3.82 days

Ionisation Energies/kJ mol^{-1}

M - M$^+$	1037
M$^+$ - M^{2+}	
M^{2+} - M^{3+}	
M^{3+} - M^{4+}	
M^{4+} - M^{5+}	
M^{5+} - M^{6+}	
M^{6+} - M^{7+}	
M^{7+} - M^{8+}	
M^{8+} - M^{9+}	
M^{9+} - M^{10+}	

Other Information

Enthalpy of Fusion/kJ mol^{-1}	2.7
Enthalpy of Vaporisation/kJ mol^{-1}	18.1

Oxidation State   Rn^0

Covalent Bonds /kJ mol^{-1}

not applicable

Rhenium   *Re*

General Information

Discovery

Rhenium was discovered by W. Noddack, I. Tacke and O. Berg in 1925 in Berlin, Germany.

Appearance

Rhenium is a silvery metal which is usually obtained as a grey powder.

Source

Rhenium does not occur free in nature or as a compound in a mineral species. It is, however, widely spread throughout the earth's crust to the extent of about 0.001 parts per million. Commercial production of rhenium is by extraction from the flue dusts of molybdenum smelters.

Uses

Rhenium is used as an additive to tungsten and molybdenum-based alloys to impart useful properties. It is widely used for filaments for mass spectrographs. It is also used as an electrical contact material as it has good wear resistance and withstands arc corrosion. Rhenium catalysts are exceptionally resistant to poisoning and are used for the hydrogenation of fine chemicals.

Biological Role

Rhenium has no known biological role.

General Information

Rhenium resists corrosion and oxidation but slowly tarnishes in moist air. It dissolves in nitric and sulphuric acids.

Physical Information

Atomic Number	75
Relative Atomic Mass ($^{12}C=12.000$)	186.2
Melting Point/K	3453
Boiling Point/K	5900
Density/kg m^{-3}	21020 (293K)
Ground State Electron Configuration	[Xe]$4f^{14}5d^56s^2$
Electron Affinity(M-M$^-$)/kJ mol^{-1}	37

Key Isotopes

nuclide	^{185}Re	^{186}Re	^{187}Re	^{188}Re
atomic mass	184.9		186.9	
natural abundance	37.4%	0%	62.6%	0%
half-life	stable	88.9 h	4×10^{10} yrs	16.7 h

Ionisation Energies/kJ mol^{-1}

M – M$^+$	760
M$^+$ – M^{2+}	1260
M^{2+} – M^{3+}	2510
M^{3+} – M^{4+}	3640
M^{4+} – M^{5+}	4900
M^{5+} – M^{6+}	6300
M^{6+} – M^{7+}	7600
M^{7+} – M^{8+}	
M^{8+} – M^{9+}	
M^{9+} – M^{10+}	

Other Information

Enthalpy of Fusion/kJ mol^{-1} 33.1
Enthalpy of Vaporisation/kJ mol^{-1} 704.25

Oxidation States

main   Re^{III}, Re^{IV}, Re^{V}
others   Re^{-III}, Re^{-I}, Re^{O}, Re^{I}, Re^{II}, Re^{VI}, Re^{VII}

Covalent Bonds /kJ mol^{-1}

not applicable

Rhodium Rh

General Information

Discovery

Rhodium was discovered by W.H. Wollaston in 1803 in London.

Appearance

Rhodium is a lustrous, silvery, hard metal.

Source

Rhodium occurs native with other platinum metals in river sands in North and South America, and in the copper-nickel sulphide ores of Ontario. Although the quantity occuring here is very small, the large amounts of nickel processed make the extraction of rhodium as a by-product commercially feasible.

Uses

The major use of rhodium is as a hardener for platinum and palladium, to produce alloys used for electrodes, furnace windings, crucibles and thermocouple elements. It is often used as an electrical contact material as it has a low resistance and is highly resistant to corrosion. Plated rhodium is exceptionally hard and is used for optical instruments. It is also used as a catalyst.

Biological Role

Rhodium has no known biological role, but is a suspected carcinogen.

General Information

Rhodium is inert to all acids but attacked by fused alkalis. It is stable in air up to 875K.

Physical information

Atomic Number	45
Relative Atomic Mass ($^{12}C=12.000$)	102.91
Melting Point/K	2239
Boiling Point/K	4000
Density/kg m^{-3}	12410 (293K)
Ground State Electron Configuration	$[Kr]4d^8 5s^1$
Electron Affinity(M-M$^-$)/kJ mol^{-1}	162

Key Isotopes

nuclide	^{103}Rh	^{105}Rh
atomic mass	102.91	
natural abundance	100%	0%
half-life	stable	35.88 h

Ionisation Energies/kJ mol^{-1}

M – M$^+$	720
M$^+$ – M^{2+}	1744
M^{2+} – M^{3+}	2997
M^{3+} – M^{4+}	4400
M^{4+} – M^{5+}	6500
M^{5+} – M^{6+}	8200
M^{6+} – M^{7+}	10100
M^{7+} – M^{8+}	12200
M^{8+} – M^{9+}	14200
M^{9+} – M^{10+}	22000

Other Information

Enthalpy of Fusion/kJ mol^{-1} 21.55
Enthalpy of Vaporisation/kJ mol^{-1} 494.3

Oxidation States

main   Rh^{III}
others   Rh^{-I}, Rh^{O}, Rh^{I}, Rh^{II}, Rh^{IV}, Rh^{V}, Rh^{VI}

Covalent Bonds /kJ mol^{-1}

not applicable

Rubidium *Rb*

General Information

Discovery

Rubidium was discovered in 1861 by R.W. Bunsen and G. Kirchoff in Heidelberg, Germany, by spectroscopic examination of the mineral lepidolite.

Appearance

Rubidium is a very soft, silvery-white metal with a lustre when cut.

Source

Rubidium is the 16th most abundant element in the earth's crust. It occurs in the minerals pollucite, carnallite, leucite and lepidolite, from which it is recovered commercially. Potassium minerals and brines also contain this element and are a further commercial source.

Uses

Rubidium is used little outside research. It is easily ionised so was considered for use in ion engines, but was found to be less effective than caesium. It has been proposed for use as a working fluid for vapour turbines and in thermoelectric generators. It is used as a photocell component and in special glasses.

Biological Role

Rubidium has no known biological role and is non-toxic. It is slightly radioactive and so has been used to locate brain tumours, as it collects in tumours but not in normal tissue.

General Information

Rubidium can be liquid at room temperature. It ignites spontaneously in air and reacts violently with water, igniting the liberated hydrogen. It forms amalgams with mercury and alloys with gold, caesium, potassium and sodium. It colours a flame yellowish-violet.

Physical Information

Atomic Number	37
Relative Atomic Mass ($^{12}C=12.000$)	85.47
Melting Point/K	312.2
Boiling Point/K	961
Density/kg m^{-3}	1532 (293K)
Ground State Electron Configuration	[Kr]$5s^1$
Electron Affinity(M-M$^-$)/kJ mol^{-1}	46.9

Key Isotopes

nuclide	^{83}Rb	^{85}Rb	^{86}Rb	^{87}Rb
atomic mass		84.91	85.91	86.91
natural abundance	0%	72.17%	0%	27.83%
half-life	83 days	stable	18.66 days	5×10^{11} yrs

Ionisation Energies/kJ mol^{-1}

M - M$^+$	403
M$^+$ - M^{2+}	2632
M^{2+} - M^{3+}	3900
M^{3+} - M^{4+}	5080
M^{4+} - M^{5+}	6850
M^{5+} - M^{6+}	8140
M^{6+} - M^{7+}	9570
M^{7+} - M^{8+}	13100
M^{8+} - M^{9+}	14800
M^{9+} - M^{10+}	26740

Other Information

Enthalpy of Fusion/kJ mol^{-1}	2.2
Enthalpy of Vaporisation/kJ mol^{-1}	75.7

Oxidation States Rb^{-I}, Rb^{I}

Covalent Bonds /kJ mol^{-1}

not applicable

Ruthenium Ru

General Information

Discovery

Ruthenium was discovered by J.A. Sniadecki in 1808 in Poland, but not recognised as an element. Klaus is generally recognised as the discoverer, as in 1844 he purified the metal from the impure oxide.

Appearance

Ruthenium is a hard, lustrous, white metal that does not tarnish at room temperature.

Source

Ruthenium is found as the free metal but also associated with other platinum metals in the mineral pentlandite, found in the USA, and pyroxinite, found in South Africa. Commercially, it is obtained from the wastes of nickel refining.

Uses

Ruthenium is one of the most effective hardeners for platinum and palladium, and is alloyed with these metals to make electrical contacts for severe wear resistance. It is also a versatile catalyst, used to split hydrogen sulphide as one example.

Biological Role

Ruthenium has no known biological role. Ruthenium (IV) oxide is highly toxic.

General Information

Ruthenium is unaffected by air, water and acids but dissolves in molten alkali, and is attacked by halogens.

Physical Information

Atomic Number	44
Relative Atomic Mass ($^{12}C=12.000$)	101.07
Melting Point/K	2583
Boiling Point/K	4173
Density/kg m^{-3}	12370 (293K)
Ground State Electron Configuration	$[Kr]4d^75s^1$
Electron Affinity(M-M$^-$)/kJ mol^{-1}	146

Key Isotopes

nuclide	^{96}Ru	^{97}Ru	^{98}Ru	^{99}Ru	^{100}Ru
atomic mass	95.91		97.91	98.91	99.90
natural abundance	5.52%	0%	1.88%	12.7%	12.6%
half-life	stable	2.88 days	stable	stable	stable

nuclide	^{101}Ru	^{102}Ru	^{103}Ru	^{104}Ru	^{106}Ru
atomic mass	100.9	101.9		103.91	
natural abundance	17%	31.6%	0%	18.7%	0%
half-life	stable	stable	39.6 days	stable	367 days

Ionisation Energies/kJ mol^{-1}

M - M$^+$	711
M$^+$ - M^{2+}	1617
M^{2+} - M^{3+}	2747
M^{3+} - M^{4+}	4500
M^{4+} - M^{5+}	6100
M^{5+} - M^{6+}	7800
M^{6+} - M^{7+}	9600
M^{7+} - M^{8+}	11500
M^{8+} - M^{9+}	18700
M^{9+} - M^{10+}	20900

Other Information

Enthalpy of Fusion/kJ mol^{-1} 23.7
Enthalpy of Vaporisation/kJ mol^{-1} 567

Oxidation States

main   Ru^{III}
others   Ru^{-II}, Ru^{O}, Ru^{I}, Ru^{II}, Ru^{IV}, Ru^{V}, Ru^{VI}, Ru^{VII}, Ru^{VIII}

Covalent Bonds /kJ mol^{-1}

not applicable

Samarium   *Sm*

General Information

Discovery
Samarium was discovered by P.E. Lecoq de Boisbaudran in 1879 in Paris.

Appearance
Samarium is a silvery-white metal with a bright sheen.

Source
Samarium is found along with other rare earth metals in several minerals, the principal ones being monazite and bastnaesite. It can be separated from the other components of the mineral by ion exchange and solvent extraction. Recently, electrochemical deposition using lithium citrate as the electrolyte and a mercury electrode has been used to separate samarium from other rare earth elements. Samarium can also be produced by reducing the oxide with barium.

Uses
Samarium is used to dope calcium fluoride crystals for use in optical lasers. It is also used in infrared absorbing glass and as a neutron absorber in nuclear reactors. In common with other rare earth elements, samarium is used in carbon arc lighting for studio lighting and projection.

Biological Role
Samarium has no known biological role, and has low toxicity.

General Information
Samarium is relatively stable in dry air but an oxide coating forms in moist air. The metal ignites in air at 150K.

Physical Information

Atomic Number	62
Relative Atomic Mass ($^{12}C=12.000$)	150.36
Melting Point/K	1350
Boiling Point/K	2064
Density/kg m^{-3}	7520 (293K)
Ground State Electron Configuration	$[Xe]4f^6 6s^2$
Electron Affinity(M-M$^-$)/kJ mol^{-1}	50

Key Isotopes

nuclide	^{144}Sm	^{146}Sm	^{147}Sm	^{148}Sm	^{149}Sm
atomic mass	143.9	145.9	146.9	147.9	148.9
natural abundance	3.1%	0%	15.1%	11.3%	13.9%
half-life	stable	7x10^{7}yrs	1.05x10^{11}yrs	12x10^{14}yrs	1x10^{15}yrs

nuclide	^{150}Sm	^{152}Sm	^{153}Sm	^{154}Sm
atomic mass	149.9	151.9		153.9
natural abundance	7.4%	26.6%	0%	22.6%
half-life	stable	stable	46.8 h	stable

Ionisation Energies/kJ mol^{-1}

M - M$^+$	543.3
M$^+$ - M^{2+}	1068
M^{2+} - M^{3+}	2260
M^{3+} - M^{4+}	3990
M^{4+} - M^{5+}	
M^{5+} - M^{6+}	
M^{6+} - M^{7+}	
M^{7+} - M^{8+}	
M^{8+} - M^{9+}	
M^{9+} - M^{10+}	

Other Information

Enthalpy of Fusion/kJ mol^{-1}	10.9
Enthalpy of Vaporisation/kJ mol^{-1}	164.8

Oxidation States

main	Sm^{III}
other	Sm^{II}

Covalent Bonds /kJ mol^{-1}

not applicable

Scandium   Sc

General Information

Discovery

Scandium was discovered by L.F. Nilson in 1879 in Uppsala, Sweden. It was, however, predicted by Mendeleev who named it ekaboron.

Appearance

Scandium is a soft, silvery-white metal, which becomes slightly tinged with yellow or pink upon exposure to the air.

Source

Scandium is the 50th most abundant element on the earth. It is very widely distributed, and occurs in minute quantities in over 800 mineral species.

In the rare mineral thortveitite, however, which is found in Scandinavia, it is the principal component.

Scandium can be recovered from thortveitite or extracted as a by-product from uranium mill tailings. Metallic scandium can also be prepared by electrolysing a eutectic melt of potassium, lithium and scandium chlorides, with electrodes of tungsten wire and a pool of molten zinc.

Uses

Scandium is not widely used. Scandium iodide is added to mercury vapour lamps to produce a highly efficient light source resembling sunlight, which is important for indoor and night-time colour television transmission. The radioactive isotope ^{46}Sc is used as a tracing agent in refinery crackers for crude oil. However, the potential for scandium is very great indeed because it is almost as light as aluminium and has a much higher melting point, so has attracted the interest of space missile designers.

Biological Role

Scandium has no known biological role, but is a suspected carcinogen.

General Information

Scandium is a much more abundant element in the sun and in certain stars than here on earth. The blue colour of beryl (the aquamarine variety) is attributed to scandium.

Physical Information

Atomic Number	21
Relative Atomic Mass ($^{12}C=12.000$)	44.956
Melting Point/K	1814
Boiling Point/K	3104
Density/kg m^{-3}	2989 (273K)
Ground State Electron Configuration	[Ar]$3d^1 4s^2$
Electron Affinity(M-M$^-$)/kJ mol^{-1}	-70

Key Isotopes

nuclide	^{44}Sc	^{45}Sc	^{46}Sc	^{47}Sc
atomic mass		44.956	45.955	
natural abundance	0%	100%	0%	0%
half-life	3.92 h	stable	83.80 days	3.34 days

Ionisation Energies/kJ mol^{-1}

M - M$^+$	631
M$^+$ - M^{2+}	1235
M^{2+} - M^{3+}	2389
M^{3+} - M^{4+}	7089
M^{4+} - M^{5+}	8844
M^{5+} - M^{6+}	10720
M^{6+} - M^{7+}	13320
M^{7+} - M^{8+}	15310
M^{8+} - M^{9+}	17369
M^{9+} - M^{10+}	21740

Other Information

Enthalpy of Fusion/kJ mol^{-1}	15.9
Enthalpy of Vaporisation/kJ mol^{-1}	376.1

Oxidation States Sc^{II}, Sc^{III}

Covalent Bonds /kJ mol^{-1}

not applicable

Selenium   Se

General Information

Discovery

Selenium was discovered by J.J.Berzelius in 1817 in Stockholm, Sweden.

Appearance

Selenium exists as several allotropic forms. The most stable variety, crystalline hexagonal selenium, is a metallic grey colour. Crystalline monoclinic selenium is deep red. Amorphous selenium is either red (in powder form) or black (in vitreous form).

Source

Most of the world's selenium is obtained from the anode muds from electrolytic copper refineries. These muds are either roasted with soda or sulphuric acid, or smelted with soda to release the selenium. Selenium is found in a few rare minerals.

Uses

Selenium exhibits both photovoltaic action (converts light to electricity) and photoconductive action (electrical resistance decreases with increased illumination). Selenium is therefore useful in photocells and solar cells. It can also convert a.c. electricity to d.c. electricity, so is extensively used in rectifiers. It is used by the glass industry, and to make stainless steel. It is also used in photocopiers.

Biological Role

Selenium is an essential trace element but is toxic in excess. It is carcinogenic and teratogenic. Hydrogen selenide and other selenium compounds are extremely toxic.

General Information

Selenium burns in air, is unaffected by water and dissolves in concentrated nitric acid and alkalis.

Physical Information

Atomic Number	34
Relative Atomic Mass ($^{12}C=12.000$)	78.96
Melting Point/K	490
Boiling Point/K	958.1
Density/kg m^{-3}	4790 (293K)
Ground State Electron Configuration	[Ar]$3d^{10}4s^{2}4p^{4}$
Electron Affinity(M-M^{-})/kJ mol^{-1}	195

Key Isotopes

nuclide	^{74}Se	^{75}Se	^{76}Se	^{77}Se
atomic mass	73.923	74.923	75.919	76.920
natural abundance	0.9%	0%	9%	7.6%
half-life	stable	120.4 days	stable	stable

nuclide	^{78}Se	^{80}Se	^{82}Se
atomic mass	77.917	79.917	81.917
natural abundance	23.5%	49.6%	9.4%
half-life	stable	stable	stable

Ionisation Energies/kJ mol^{-1}

M - M^{+}	940.9
M^{+} - M^{2+}	2044
M^{2+} - M^{3+}	2974
M^{3+} - M^{4+}	4144
M^{4+} - M^{5+}	6590
M^{5+} - M^{6+}	7883
M^{6+} - M^{7+}	14990
M^{7+} - M^{8+}	19500
M^{8+} - M^{9+}	23300
M^{9+} - M^{10+}	27200

Other Information

Enthalpy of Fusion/kJ mol^{-1}	5.1
Enthalpy of Vaporisation/kJ mol^{-1}	90

Oxidation States

main	Se^{IV}, Se^{VI}
others	Se^{-II}, Se^{I}, Se^{II}

Covalent Bonds /kJ mol^{-1}

Se-H	305
Se-C	245
Se-O	343
Se-F	285
Se-Cl	245
Se-Se	330

Silicon Si

General Information

Discovery

J.J.Berzelius is credited with the discovery of silicon in 1824 in Stockholm, Sweden. However, Gay Lussac and Thenard probably prepared impure amorphous silicon in 1811. Deville prepared the second allotropic form of silicon, crystalline silicon, in 1854.

Appearance

Amorphous silicon is a brown powder, and crystalline silicon is a grey colour with a metallic lustre.

Source

Silicon makes up 25.7% of the earth's crust by mass and is the second most abundant element (oxygen is the first). It does not occur free in nature but occurs chiefly as the oxide and as silicates. The oxide includes sand, quartz, rock crystal, amethyst, agate, flint and opal. The silicate form includes asbestos, granite hornblende, feldspar, clay and mica.

Silicon is prepared commercially by electrolysis with carbon electrodes of a mixture of silica and carbon. Silicon is used extensively in solid-state devices and for this hyperpure silicon is required. This is prepared by thermal decompositon of ultra-pure trichlorosilane.

Uses

Silicon is one of the most useful elements to mankind. Sand and clay, which both contain silicon, are used to make concrete and cement. Sand is also the principal ingredient of glass, which has thousands of uses. Silicon is a component of steel, and silicon carbides are important abrasives and also used in lasers. Silicon is present in pottery and enamels, and in high-temperature materials.

However, increasing use for silicon is now being found in micro-electronic devices. The silicon is usually doped with boron, gallium, phosphorus or arsenic for use in transistors, solar cells, rectifiers and other instruments.

Biological Role

Silicon is essential to plant and animal life. It is non-toxic but some silicates, such as asbestos, are carcinogenic. Some workers such as miners and stonecutters who are exposed to siliceous dust often develop a serious lung disease called silicosis.

General Information (continued)

Silicon is relatively inert. It is attacked by halogens and dilute alkali, but is not attacked by acids except hydrofluoric.

Silicones are important products of silicon, prepared by hydrolysing a silicon organic chloride. Hydrolysis and condensation of substituted chlorosilanes can be used to produce a great number of polymers known as silicones, ranging from liquids to hard, glasslike solids with many useful properties.

Physical Information

Atomic Number	14
Relative Atomic Mass ($^{12}C=12.000$)	28.086
Melting Point/K	1683
Boiling Point/K	2628
Density/kg m^{-3}	2329 (293K)
Ground State Electron Configuration	[Ne]$3s^2 3p^2$
Electron Affinity(M-M$^-$)/kJ mol^{-1}	135

Key Isotopes

nuclide	^{28}Si	^{29}Si	^{30}Si	^{32}Si
atomic mass	27.977	28.976	29.974	31.974
natural abundance	92.23%	4.67%	3.10%	0%
half-life	stable	stable	stable	650 yrs

Ionisation Energies/kJ mol^{-1}

M - M$^+$	786.5
M$^+$ - M^{2+}	1577.1
M^{2+} - M^{3+}	3231.4
M^{3+} - M^{4+}	4355.5
M^{4+} - M^{5+}	16091
M^{5+} - M^{6+}	19784
M^{6+} - M^{7+}	23786
M^{7+} - M^{8+}	29252
M^{8+} - M^{9+}	33876
M^{9+} - M^{10+}	38732

Other Information

Enthalpy of Fusion/kJ mol^{-1}	39.6
Enthalpy of Vaporisation/kJ mol^{-1}	383.3

Oxidation States   Si^{II}, Si^{IV}

Covalent Bonds /kJ mol^{-1}

Si-H	326
Si-C	301
Si-O	486
Si-F	582
Si-Cl	391
Si-Si	226

Silver   *Ag*

General Information

Discovery

Silver was known to ancient civilisations, and evidence indicates that man learned to separate silver from lead in 3000 B.C.

Appearance

Silver has a brilliant white metallic lustre, with a characteristic sheen.

Source

Silver occurs native in ores such as argentite and horn silver, but the principal sources are lead, lead-zinc, copper, gold and copper-nickel ores. Canada and the USA are the main silver producers in the Western hemisphere. The metal is either recovered from the ore, or during the electrolytic refining of copper.

Uses

Sterling silver contains 92.5% silver, the remainder being copper or some other metal, and is used for jewellery and silverware where appearance is important. About 30% of silver produced is used in the photographic industry, mostly as silver (I) nitrate. Silver is used in dental alloys, solder and brazing alloys, electrical contacts and batteries. Silver paints are used for making printed circuits. The metal is used to make mirrors as it is the best reflector of visible light known, although it does tarnish with time.

Biological Role

Silver has no known biological role, although it is a suspected carcinogen. Silver compounds can be absorbed in the circulatory system and reduced silver deposited in various organs. This results in greyish pigmentation of the skin and mucous membranes, known as argyria. Silver has germicidal effects - it can kill lower organisms quite effectively.

General Information

Silver is stable to water and oxygen but is attacked by sulphur compounds in air to form a black sulphide layer. It dissolves in sulphuric and nitric acids. It is a little harder than gold and is extremely ductile and malleable. Pure silver has the highest electrical and thermal conductivity of all metals, and has the lowest contact resistance.

Physical Information

Atomic Number	47
Relative Atomic Mass ($^{12}C=12.000$)	107.87
Melting Point/K	1235.08
Boiling Point/K	2485
Density/kg m^{-3}	10500 (293K)
Ground State Electron Configuration	[Kr]$4d^{10}5s^1$
Electron Affinity(M-M$^-$)/kJ mol^{-1}	125.7

Key Isotopes

nuclide	^{107}Ag	^{109}Ag	^{111}Ag
atomic mass	106.911	108.90	
natural abundance	51.83%	48.17%	0%
half-life	stable	stable	7.5 days

Ionisation Energies/kJ mol^{-1}

M - M$^+$	731
M$^+$ - M^{2+}	2073
M^{2+} - M^{3+}	3361
M^{3+} - M^{4+}	5000
M^{4+} - M^{5+}	6700
M^{5+} - M^{6+}	8600
M^{6+} - M^{7+}	11200
M^{7+} - M^{8+}	13400
M^{8+} - M^{9+}	15600
M^{9+} - M^{10+}	18000

Other Information

Enthalpy of Fusion/kJ mol^{-1}	11.3
Enthalpy of Vaporisation/kJ mol^{-1}	257.7

Oxidation States

main	Ag^I
others	Ag^O, Ag^{II}, Ag^{III}

Covalent Bonds /kJ mol^{-1}

not applicable

Sodium   *Na*

General Information

Discovery

Sodium was isolated by Sir Humphry Davy in 1807 in London, by the electrolysis of caustic soda.

Appearance

Sodium is a soft, silvery-white metal which is generally stored in paraffin, as it oxidises rapidly when cut.

Source

Sodium is the sixth most abundant element on earth, and comprises 2.6% of the earth's crust. The most common compound is sodium chloride, but it also occurs in many minerals among which are cryolite, zeolite and sodalite. It is never found free in nature, due to its great reactivity. It is obtained commercially by the electrolysis of dry fused sodium chloride.

Uses

Metallic sodium is used in the manufacture of sodamide and esters, and in the preparation of certain organic compounds. Other uses of the metal include descaling and purifying metals, and alloy formation. One alloy of sodium with potassium is an important heat transfer agent. Sodium compounds are important in several industries, including paper, glass, soap, textile, petroleum and metal. Salt is also of vital nutritional importance.

Biological Role

Sodium is essential to all animals, and this has been recognised since prehistoric times. Although it is considered non-toxic, too much salt in the diet has been linked to high blood pressure under certain circumstances.

General Information

Sodium is very reactive, and should be handled with care. It floats on water, decomposing it with the evolution of hydrogen and the formation of sodium hydroxide. It may or may not ignite spontaneously on the water, depending on the amount of metal exposed to the water. It normally does not ignite in air at temperatures below 115K.

Physical Information

Atomic Number	11
Relative Atomic Mass ($^{12}C=12.000$)	22.990
Melting Point/K	370.96
Boiling Point/K	1156.1
Density/kg m^{-3}	971 (273K)
Ground State Electron Configuration	[Ne]3s^1
Electron Affinity(M-M$^-$)/kJ mol^{-1}	21

Key Isotopes

nuclide	^{22}Na	^{23}Na	^{24}Na
atomic mass	21.994	22.990	23.991
natural abundance	0%	100%	0%
half-life	2.602 yrs	stable	15.0 h

Ionisation Energies/kJ mol^{-1}

M - M$^+$	495.8
M$^+$ - M^{2+}	4562.4
M^{2+} - M^{3+}	6912
M^{3+} - M^{4+}	9543
M^{4+} - M^{5+}	13353
M^{5+} - M^{6+}	16610
M^{6+} - M^{7+}	20114
M^{7+} - M^{8+}	25490
M^{8+} - M^{9+}	28933
M^{9+} - M^{10+}	141360

Other Information

Enthalpy of Fusion/kJ mol^{-1}	2.64
Enthalpy of Vaporisation/kJ mol^{-1}	99.2

Oxidation States

main	Na^{I}
others	Na^{-I}

Covalent Bonds /kJ mol^{-1}

not applicable

Strontium Sr

General Information

Discovery

Strontium was isolated by Sir Humphry Davy in 1808 in London, but recognised as an element by A Crawford in 1790.

Appearance

Strontium is a silvery-white, soft metal which rapidly forms the yellowish colour of the oxide when cut.

Source

Strontium is found mainly in the minerals celestite and strontianite. It can be prepared by electrolysis of the fused chloride with potassium chloride, or by reducing strontium oxide with aluminium.

Uses

Strontium is mainly used for producing glass for colour television sets. It is also used in producing ferrite magnets and refining zinc. One of the radioactive isotopes of strontium, ^{90}Sr, is a product of nuclear fallout and presents a health problem. It has a half-life of 28 years. It is absorbed by bone tissue instead of calcium and can destroy bone marrow and cause cancer. However, it is also a useful isotope as it is one of the best high-energy beta-emitters known.

Biological Role

Strontium has no known biological role, and it is non-toxic. It replaces and mimics calcium.

General Information

Strontium will burn in air and reacts with water more vigorously than calcium. It is usually kept under paraffin to prevent oxidation.

Physical Information

Atomic Number	38
Relative Atomic Mass ($^{12}C=12.000$)	87.62
Melting Point/K	1042
Boiling Point/K	1657
Density/kg m^{-3}	2540 (293K)
Ground State Electron Configuration	[Kr]$5s^2$
Electron Affinity(M-M$^-$)/kJ mol^{-1}	-146

Key Isotopes

nuclide	^{82}Sr	^{84}Sr	^{85}Sr	^{86}Sr
atomic mass		83.91	84.91	85.91
natural abundance	0%	0.56%	0%	9.86%
half-life	25 days	stable	64 days	stable

nuclide	^{87}Sr	^{88}Sr	^{89}Sr	^{90}Sr
atomic mass	86.91	87.91	88.91	89.91
natural abundance	7%	82.58%	0%	0%
half-life	stable	stable	52.7 days	28.1 yrs

Ionisation Energies/kJ mol^{-1}

M - M$^+$	549.5
M$^+$ - M^{2+}	1064.2
M^{2+} - M^{3+}	4210
M^{3+} - M^{4+}	5500
M^{4+} - M^{5+}	6910
M^{5+} - M^{6+}	8760
M^{6+} - M^{7+}	10200
M^{7+} - M^{8+}	11800
M^{8+} - M^{9+}	15600
M^{9+} - M^{10+}	17100

Other Information

Enthalpy of Fusion/kJ mol^{-1}	9.16
Enthalpy of Vaporisation/kJ mol^{-1}	154.4

Oxidation States   Sr^{II}

Covalent Bonds /kJ mol^{-1}

not applicable

Sulphur S

General Information

Discovery

Sulphur was known to ancient civilisations, and referred to in Genesis as brimstone.

Appearance

Sulphur exists as several allotropes, of which orthorhombic sulphur is the most stable. It is a pale yellow, brittle, odourless solid.

Source

Sulphur is widely distributed in nature as iron pyrites, galena, gypsum, Epsom salts and many other minerals. It is commercially recovered from wells sunk into the salt domes along the Gulf Coast of the USA, and from the Alberta gas fields. It is also mined in Poland. The Frasch Process is used to force heated water into the wells to melt the sulphur, which can then be recovered chemically. Sulphur can also be recovered from natural gas and crude oil by conversion into hydrogen sulphide, from which sulphur is liberated.

Uses

Sulphur is used in the vulcanisation of black rubber, as a fungicide and in black gunpowder. Most, however, is used in the production of sulphuric acid, which is the most important chemical manufactured by western civilisations.

Biological Role

Sulphur is essential to life as a component of fats, body fluids and bones. It is non-toxic as the element and in the form of the sulphate, but carbon disulphide, hydrogen sulphide and sulphur dioxide are all toxic, especially hydrogen sulphide which can cause death by respiratory paralysis.

General Information

Sulphur occurs in several allotropic forms whether in the liquid, solid or gaseous state. Amorphous or 'plastic' sulphur is obtained by fast cooling of the crystalline form, and is thought to have a helical structure with eight atoms per spiral. Crystalline sulphur is made up of rings, each containing eight sulphur atoms.

Physical Information

Atomic Number	16
Relative Atomic Mass ($^{12}C=12.000$)	32.066
Melting Point/K	386.0
Boiling Point/K	717.824
Density/kg m^{-3}	2070 (293K)
Ground State Electron Configuration	[Ne]$3s^2 3p^4$
Electron Affinity(M-M$^-$)/kJ mol^{-1}	200.4

Key Isotopes

nuclide	^{32}S	^{33}S	^{34}S	^{35}S	^{36}S
atomic mass	31.972	32.971	33.968	34.969	35.967
natural abundance	95.02%	0.75%	4.21%	0%	0.02%
half-life	stable	stable	stable	87.9 days	stable

Ionisation Energies/kJ mol^{-1}

M - M$^+$	999.6
M$^+$ - M^{2+}	2251
M^{2+} - M^{3+}	3361
M^{3+} - M^{4+}	4564
M^{4+} - M^{5+}	7013
M^{5+} - M^{6+}	8495
M^{6+} - M^{7+}	27106
M^{7+} - M^{8+}	31669
M^{8+} - M^{9+}	36578
M^{9+} - M^{10+}	43138

Other Information

Enthalpy of Fusion/kJ mol^{-1}	1.23
Enthalpy of Vaporisation/kJ mol^{-1}	9.62

Oxidation States

main	S^{VI}
others	S^{-II}, S^{-I}, S^{O}, S^{I}, S^{II}, S^{III}, S^{IV}, S^{V}

Covalent Bonds /kJ mol^{-1}

S-H	347
S-C	272
S=C	476
S-O	265
S=O	525
S-F	328
S-Cl	255
S-S	226

Tantalum   Ta

General Information

Discovery

Tantalum was discovered by A.G. Ekeberg in 1802 in Uppsala, Sweden, but many chemists thought that tantalum and niobium were identical elements until Rose (in 1844) and Marignac (in 1866) showed that niobic and tantalic acids were different.

Appearance

Tantalum is a shiny, grey metal which is soft when pure.

Source

Tantalum occurs principally in the mineral columbite-tantalite, found in many places including Australia, Canada and Africa. Separation of tantalum from niobium requires several complicated steps. It is obtained commercially as a by-product of tin extraction.

Uses

Tantalum causes no immune response in mammals, so has found wide use in the making of surgical appliances. It can replace bone, for example in skull plates; as foil or wire it connects torn nerves; as woven gauze it binds abdominal muscle. Tantalum has also been used to make a variety of alloys.

Biological Role

Tantalum has no known biological role, and is non-toxic.

General Information

Tantalum is very corrosion resistant due to the formation of an oxide film, but is attacked by hydrogen fluoride and fused alkalis. It has a melting point exceeded only by tungsten and rhenium.

Physical Information

Atomic Number	73
Relative Atomic Mass ($^{12}C=12.000$)	180.95
Melting Point/K	3269
Boiling Point/K	5698
Density/kg m^{-3}	16654 (293K)
Ground State Electron Configuration	[Xe]$4f^{14}5d^{3}6s^{2}$
Electron Affinity(M-M$^-$)/kJ mol^{-1}	14

Key Isotopes

nuclide	^{180}Ta	^{181}Ta	^{182}Ta
atomic mass	179.9	180.9	
natural abundance	0.012%	99.99%	0%
half-life	1×10^{12} yrs	stable	115.1 days

Ionisation Energies/kJ mol^{-1}

M − M$^+$	761
M$^+$ − M^{2+}	1500
M^{2+} − M^{3+}	2100
M^{3+} − M^{4+}	3200
M^{4+} − M^{5+}	4300
M^{5+} − M^{6+}	
M^{6+} − M^{7+}	
M^{7+} − M^{8+}	
M^{8+} − M^{9+}	
M^{9+} − M^{10+}	

Other Information

Enthalpy of Fusion/kJ mol^{-1} 31.4
Enthalpy of Vaporisation/kJ mol^{-1} 758.2

Oxidation States
main   Ta^{V}
others   Ta^{-III}, Ta^{-I}, Ta^{I}, Ta^{II}, Ta^{III}, Ta^{IV}

Covalent Bonds /kJ mol^{-1}
not applicable

Technetium   *Tc*

General Information

Discovery

Technetium was discovered by C. Perrier and E.G. Segre in 1937 in Palermo, Italy. It was the first element to be produced artificially.

Appearance

Technetium is a silvery-grey metal that tarnishes slowly in moist air. It is usually obtained as a grey powder.

Source

The metal is produced in tonne quantities from the fission products of uranium nuclear fuel.

Uses

Technetium is a remarkable corrosion inhibitor for steel, and can protect steel by the addition of very small amounts. This use is limited to closed systems as technetium is radioactive.

Biological Role

Technetium has no known biological role. It is a radioactive element.

General Information

Technetium is an excellent superconductor at 11K and below. It resists oxidation, burns in oxygen and dissolves in nitric and sulphuric acids.

Physical Information

Atomic Number	43
Relative Atomic Mass ($^{12}C=12.000$)	98.91
Melting Point/K	2445
Boiling Point/K	5150
Density/kg m^{-3}	11500 (293K)
Ground State Electron Configuration	[Kr]4d^{5}5s^2
Electron Affinity(M-M$^-$)/kJ mol^{-1}	96

Key Isotopes

nuclide	^{97}Tc	^{98}Tc	^{99}Tc
atomic mass		97.911	98.90
natural abundance	0%	0%	0%
half-life	2.6×10^6 yrs	1.5×10^6 yrs	2.12×10^5 yrs

Ionisation Energies/kJ mol^{-1}

M - M$^+$	702
M$^+$ - M^{2+}	1472
M^{2+} - M^{3+}	2850
M^{3+} - M^{4+}	4100
M^{4+} - M^{5+}	5700
M^{5+} - M^{6+}	7300
M^{6+} - M^{7+}	9100
M^{7+} - M^{8+}	15600
M^{8+} - M^{9+}	17800
M^{9+} - M^{10+}	19900

Other Information

Enthalpy of Fusion/kJ mol^{-1} 23.81
Enthalpy of Vaporisation/kJ mol^{-1} 585.22

Oxidation States

main   Tc^{IV}, Tc^V, Tc^{VII}
others   Tc^{-I}, Tc^O, Tc^{VI}

Covalent Bonds /kJ mol^{-1}

not applicable

Tellurium   Te

General Information

Discovery

Tellurium was discovered by Baron Muller von Reichenstein in 1783 in Sibiu, Romania. Klaproth isolated the element and named it in 1798.

Appearance

Crystalline tellurium is a silvery-white colour with a metallic lustre, but is most often seen as the grey, powdery amorphous form.

Source

Tellurium is present in the earth's crust only in 0.001 parts per million. It is obtained commercially from the anode muds produced during the electrolytic refining of copper.

Uses

Tellurium is used in alloys, mostly with copper and stainless steel, to improve their machinability. When added to lead it decreases the corrosive action of sulphuric acid on lead and improves its strength and hardness. Tellurium is also used in ceramics. Finally, it can be doped with silver, gold, copper or tin in semiconductor applications.

Biological Role

Tellurium has no known biological role. It is very toxic and teratogenic. Workmen exposed to very small quantities of tellurium in the air develop "tellurium breath", which has a garlic-like odour.

General Information

Tellurium burns in air or oxygen with a greenish-blue flame, forming tellurium (IV) oxide. It is unaffected by water or hydrochloric acid, but dissolves in nitric acid. Tellurium is a p-type semiconductor, and its conductivity increases slightly with exposure to light. Molten tellurium corrodes iron, copper and stainless steel.

Physical Information

Atomic Number	52
Relative Atomic Mass ($^{12}C=12.000$)	127.6
Melting Point/K	722.7
Boiling Point/K	1263
Density/kg m^{-3}	6240 (293K)
Ground State Electron Configuration	[Kr]$4d^{10}5s^25p^4$
Electron Affinity(M-M$^-$)/kJ mol^{-1}	190.2

Key Isotopes

nuclide	^{120}Te	^{122}Te	^{123}Te	^{124}Te	^{125}Te
atomic mass	119.9	121.9	122.9	123.9	124.9
natural abundance	0.096%	2.6%	0.908%	4.816%	7.18%
half-life	stable	stable	1.2x10^{13}yrs	stable	stable

nuclide	^{126}Te	^{127}Te	^{128}Te	^{130}Te
atomic mass	125.9		127.9	129.9
natural abundance	18.95%	0%	31.69%	33.8%
half-life	stable	9.4 h	stable	stable

Ionisation Energies/kJ mol^{-1}

M - M$^+$	869.2
M$^+$ - M^{2+}	1795
M^{2+} - M^{3+}	2698
M^{3+} - M^{4+}	3610
M^{4+} - M^{5+}	5668
M^{5+} - M^{6+}	6822
M^{6+} - M^{7+}	13200
M^{7+} - M^{8+}	15800
M^{8+} - M^{9+}	18500
M^{9+} - M^{10+}	21200

Other information

Enthalpy of Fusion/kJ mol^{-1}	13.5
Enthalpy of Vaporisation/kJ mol^{-1}	104.6

Oxidation States

main	Te^{IV}
others	Te^{-II}, Te^{-I}, Te^{O}, Te^{II}, Te^{V}, Te^{VI}

Covalent Bonds /kJ mol^{-1}

Te-H	240
Te-O	268
Te-F	335
Te-Cl	251
Te-Te	235

Terbium   *Tb*

General Information

Discovery

Terbium was discovered by C.G. Mosander in 1843 in Stockholm, Sweden.

Appearance

Terbium is a silver-grey metal, malleable, ductile, and soft enough to be cut with a knife.

Source

Terbium can be recovered from the mineral monazite by ion exchange and solvent extraction, and from euxenite, a complex oxide containing 1% or more of terbium. It is usually produced commercially by reducing the anhydrous fluoride or chloride with calcium metal.

Uses

Terbium is used to dope calcium fluoride, calcium tungstate and strontium molybdate, all used in solid-state devices. Terbium salts are used in laser devices, but otherwise this element is not widely used.

Biological Role

Terbium has no known biological role, and has low toxicity.

General Information

Terbium is slowly oxidised by air, and reacts with cold water.

Physical Information

Atomic Number	65
Relative Atomic Mass ($^{12}C=12.000$)	158.92
Melting Point/K	1629
Boiling Point/K	3396
Density/kg m^{-3}	8229 (293K)
Ground State Electron Configuration	[Xe]4f^{9}6s^2
Electron Affinity(M-M$^-$)/kJ mol^{-1}	50

Key Isotopes

nuclide	^{159}Tb	^{160}Tb
atomic mass	158.9	
natural abundance	100%	0%
half-life	stable	72.1 days

Ionisation Energies/kJ mol^{-1}

M - M$^+$	564.6
M$^+$ - M^{2+}	1112
M^{2+} - M^{3+}	2114
M^{3+} - M^{4+}	3839
M^{4+} - M^{5+}	
M^{5+} - M^{6+}	
M^{6+} - M^{7+}	
M^{7+} - M^{8+}	
M^{8+} - M^{9+}	
M^{9+} - M^{10+}	

Other Information

Enthalpy of Fusion/kJ mol^{-1} 16.3
Enthalpy of Vaporisation/kJ mol^{-1} 391

Oxidation States Tb^{III}, Tb^{IV}

Covalent Bonds /kJ mol^{-1}

not applicable

Thallium   Tl

General Information

Discovery

Thallium was discovered spectroscopically by W. Crookes in 1861 in London.

Appearance

Thallium is a soft, silvery metal, but it soon develops a bluish-grey tinge as the oxide forms if it is exposed to the air.

Source

Thallium is found in several ores, one of which is pyrites, used in the production of sulphuric acid. The commercial source of thallium is as a by-product of pyrites roasting in sulphuric acid production. It can also be obtained from the smelting of lead and zinc ores. Thallium is also present in manganese nodules found on the ocean floor.

Uses

The use of thallium is limited as it is a toxic element. Thallium sulphate is employed as a rodent killer - it is odourless and tasteless - but household use of this poison has been prohibited in the USA. Thallium oxide is used to produce glasses with a high index of refraction, and also low melting glasses which become fluid at about 125K.

Biological Role

Thallium has no known biological role. It is very toxic and teratogenic. Contact of the metal with the skin is dangerous, and there is evidence that the vapour is both teratogenic and carcinogenic.

General Information

Thallium is soft, malleable and can be cut with a knife. It tarnishes readily in moist air and reacts with steam to form the hydroxide. It is attacked by all acids, most rapidly nitric acid.

Physical Information

Atomic Number	81
Relative Atomic Mass ($^{12}C=12.000$)	204.38
Melting Point/K	576.7
Boiling Point/K	1730
Density/kg m^{-3}	11850 (293K)
Ground State Electron Configuration	[Xe]4f^{14}5d^{10}6s^{2}6p^1
Electron Affinity(M-M$^-$)/kJ mol^{-1}	30

Key Isotopes

nuclide	^{203}Tl	^{204}Tl	^{205}Tl	^{208}Tl
atomic mass	202.97		204.97	
natural abundance	29.52%	0%	70.48%	trace
half-life	stable	3.81 yrs	stable	3.1 mins

Ionisation Energies/kJ mol^{-1}

M - M$^+$	589.3
M$^+$ - M^{2+}	1971
M^{2+} - M^{3+}	2878
M^{3+} - M^{4+}	4900
M^{4+} - M^{5+}	6100
M^{5+} - M^{6+}	8300
M^{6+} - M^{7+}	9500
M^{7+} - M^{8+}	11300
M^{8+} - M^{9+}	14000
M^{9+} - M^{10+}	16000

Other Information

Enthalpy of Fusion/kJ mol^{-1}	4.31
Enthalpy of Vaporisation/kJ mol^{-1}	166.1

Oxidation States

main	Tl^I
others	Tl^{III}

Covalent Bonds /kJ mol^{-1}

Tl(I)-H	185
Tl(III)-C	125
Tl(III)-O	375
Tl(III)-F	460
Tl(III)-Cl	368
Tl-Tl	63

Thorium   *Th*

General Information

Discovery

Thorium was discovered by J.J. Berzelius in 1815 in Stockholm, Sweden.

Appearance

Pure thorium is a radioactive silvery-white metal which retains its lustre for several months. When contaminated with the oxide, thorium slowly tarnishes in air, becoming first grey and then black.

Source

Thorium is found in large deposits in the USA and elsewhere, but these have not been exploited as a source of the element.

Several methods are used to produce the metal, such as reducing thorium oxide with calcium and by the electrolysis of anhydrous thorium chloride.

Uses

The principal use of thorium is in the Welsbach mantle, which consists of thorium oxide amongst other compounds. This type of mantle glows with a dazzling flame when heated by gas, so are used in portable gas lights. Thorium is also an important alloying agent in magnesium as it imparts greater strength and creep resistance at high temperatures.

Thorium can be used as a source of nuclear power. It is about three times as abundant as uranium and about as abundant as lead, and there is probably more energy avaiable from thorium than both uranium and fossil fuels. However, although work has been done in developing thorium cycle convertor-reactor systems, it will be many years before such a system is operative - if at all.

Biological Role

Thorium has no known biological role. It is toxic due to its radioactivity.

General Information

Pure thorium is soft and very ductile, and has one of the highest melting points of all elements. It is slowly attacked by water and acids. Powdered thorium metal is often pyrophoric. Thorium turnings ignite when heated in air and burn with a brilliant white light.

Physical Information

Atomic Number	90
Relative Atomic Mass ($^{12}C=12.000$)	232.04
Melting Point/K	2023
Boiling Point/K	5060
Density/kg m^{-3}	11720 (293K)
Ground State Electron Configuration	[Rn]$6d^2 7s^2$

Key Isotopes

nuclide	^{228}Th	^{229}Th	^{230}Th
atomic mass	228.03	229.03	230.03
natural abundance	trace	0%	trace
half-life	1.9 yrs	7340 yrs	8×10^4 yrs

nuclide	^{231}Th	^{232}Th	^{234}Th
atomic mass	231.03	232.04	234.04
natural abundance	trace	100%	trace
half-life	25.5 h	1.41×10^{10} yrs	24.1 days

Ionisation Energies/kJ mol^{-1}

M - M$^+$	587
M$^+$ - M^{2+}	1110
M^{2+} - M^{3+}	1978
M^{3+} - M^{4+}	2780
M^{4+} - M^{5+}	
M^{5+} - M^{6+}	
M^{6+} - M^{7+}	
M^{7+} - M^{8+}	
M^{8+} - M^{9+}	
M^{9+} - M^{10+}	

Other Information

Enthalpy of Fusion/kJ mol^{-1}	19.2
Enthalpy of Vaporisation/kJ mol^{-1}	513.7

Oxidation States

main	Th^{IV}
others	Th^{II}, Th^{III}

Covalent Bonds /kJ mol^{-1}

not applicable

Thulium   *Tm*

General Information

Discovery

Thulium was discovered by P.T. Cleve in 1879 in Uppsala, Sweden.

Appearance

Thulium is a silvery metal with a bright lustre.

Source

Thulium is found principally in the mineral monazite, from which it is extracted by ion exchange and solvent extraction. It can also be isolated by reduction of the anhydrous fluoride with calcium metal, or reduction of the oxide with lanthanum metal.

Uses

When irradiated in a nuclear reactor, thulium produces an isotope that emits X-rays. A "button" of this isotope is used to make a light-weight, portable X-ray machine for medical use. The "hot" thulium is replaced every few months. Otherwise this element is little used.

Biological Role

Thulium has no known biological role, and is non-toxic.

General Information

Thulium tarnishes in air and reacts with water. It is soft, malleable and ductile, and can be cut with a knife.

Physical Information

Atomic Number	69
Relative Atomic Mass ($^{12}C=12.000$)	168.93
Melting Point/K	1818
Boiling Point/K	2220
Density/kg m^{-3}	9321 (293K)
Ground State Electron Configuration	[Xe]4f^{13}6s^2
Electron Affinity(M-M$^-$)/kJ mol^{-1}	50

Key Isotopes

nuclide	^{169}Tm	^{170}Tm
atomic mass	168.9	
natural abundance	100%	0%
half-life	stable	134 days

Ionisation Energies/kJ mol^{-1}

M − M$^+$	596.7
M$^+$ − M^{2+}	1163
M^{2+} − M^{3+}	2285
M^{3+} − M^{4+}	4119
M^{4+} − M^{5+}	
M^{5+} − M^{6+}	
M^{6+} − M^{7+}	
M^{7+} − M^{8+}	
M^{8+} − M^{9+}	
M^{9+} − M^{10+}	

Other Information

Enthalpy of Fusion/kJ mol^{-1} 18.4
Enthalpy of Vaporisation/kJ mol^{-1} 247

Oxidation States

main   Tm^{III}
others   Tm^{II}

Covalent Bonds /kJ mol^{-1}

not applicable

Tin   Sn

General Information

Discovery

Tin was known to ancient civilisations.

Appearance

Tin is a silvery-white metal. It is soft, pliable and has a highly crystalline structure.

Source

Tin is found mainly in the ore cassiterite, which is found in Malaya, Bolivia, Indonesia, Thailand and Nigeria. It is obtained commercially by reducing the ore with coal in a reverberatory furnace.

Uses

Tin has many uses. It takes a high polish and is used to coat other metals to prevent corrosion, such as in tin cans which are made of tin-coated steel. Alloys of tin are important, such as soft solder, pewter, bronze and phosphor bronze. The most important tin salt used is tin (II) chloride which is used as a reducing agent and as a mordant. Tin salts sprayed onto glass are used to produce electrically conductive coatings. Most window glass is made by floating molten glass on molten tin to produce a flat surface. Recently, a tin-niobium alloy that is superconductive at very low temperatures has attracted interest.

Biological Role

Tin is non-toxic. Trialkyl and triaryl tin compounds are used as biocides and must be handled with care.

General Information

Tin is unreactive to water and oxygen, as it is protected by an oxide film. It dissolves in acids and bases. When heated in air tin forms tin (IV) oxide which is feebly acidic.

When a tin bar is broken, a "tin cry" is heard due to the breaking of the tin crystals.

Tin has two allotropic forms. On warming, grey tin, with a cubic structure, changes into white tin, the ordinary form of the metal.

Physical Information

Atomic Number	50
Relative Atomic Mass ($^{12}C=12.000$)	118.71
Melting Point/K	505.1
Boiling Point/K	2543
Density/kg m^{-3}	7310 (293K)
Ground State Electron Configuration	[Kr]4d^{10}5s^{2}5p^2
Electron Affinity(M-M$^-$)/kJ mol^{-1}	121

Key Isotopes

nuclide	^{112}Sn	^{113}Sn	^{114}Sn	^{115}Sn
atomic mass	111.91		113.9	114.9
natural abundance	1%	0%	0.7%	0.4%
half-life	stable	115 days	stable	stable

nuclide	^{116}Sn	^{117}Sn	^{118}Sn	^{119}Sn
atomic mass	115.9	116.9	117.9	118.9
natural abundance	14.7%	7.7%	24.3%	8.6%
half-life	stable	stable	stable	stable

nuclide	^{120}Sn	^{121}Sn	^{122}Sn	^{124}Sn
atomic mass	119.9		121.9	123.9
natural abundance	32.4%	0%	4.6%	5.6%
half-life	stable	27.5 h	stable	stable

Ionisation Energies/kJ mol^{-1}	
M − M$^+$	708.6
M$^+$ − M^{2+}	1411.8
M^{2+} − M^{3+}	2943
M^{3+} − M^{4+}	3930.2
M^{4+} − M^{5+}	6974
M^{5+} − M^{6+}	9900
M^{6+} − M^{7+}	12200
M^{7+} − M^{8+}	14600
M^{8+} − M^{9+}	17000
M^{9+} − M^{10+}	20600

Other Information

Enthalpy of Fusion/kJ mol^{-1}	7.2
Enthalpy of Vaporisation/kJ mol^{-1}	296.2

Oxidation States Sn^{II}, Sn^{IV}

Covalent Bonds /kJ mol^{-1}

Sn-H	314
Sn-C	225
Sn(II)-O	557
Sn(IV)-F	322
Sn(IV)-Cl	315
Sn-Sn	195

Titanium   *Ti*

General Information

Discovery

Titanium was discovered by the Rev. W. Gregor in 1791 in Creed, Cornwall, and named by M.H. Klaproth in 1795 in Berlin. However, the pure metal was not made until 1910 by Hunter, who heated titanium (IV) chloride with sodium in a steel bomb.

Appearance

Titanium is a hard, lustrous, silvery metal.

Source

Titanium is the ninth most abundant element on earth. It is almost always present in igneous rocks and the sediments derived from them. It occurs in the minerals rutile, ilmenite, and sphene, and is present in titanates and many iron ores.

Titanium is produced commercially by reducing titanium tetrachloride with magnesium.

The major use of this element is as titanium oxide. This is produced commercially by either the Sulphate Process or the Chloride Process, both of which prepare titanium oxide from the mineral ilmenite.

Uses

Titanium is as strong as steel but much lighter. It is therefore important as an alloying agent with many metals including aluminium, molybdenum and iron. These alloys are principally used in aircraft and missiles as they are materials which are light yet can withstand extremes of temperature. Titanium also has potential use in desalination plants which convert sea water to fresh water. The metal has excellent resistance to sea water and so is used to protect the hulls of ships, and other parts exposed to sea water.

However, the largest use of titanium is in the form of titanium oxide, which is extensively used in both house paint and artists' paint. This paint is also a good reflector of infra-red radiation and so is used in solar observatories where heat causes poor visibility.

Biological Role

Titanium has no known biological role, and is non-toxic. It can have a stimulatory effect, and is a suspected carcinogen.

General Information

Titanium burns in air and is the only element that burns in nitrogen. It is ductile only in an oxygen-free atmosphere. It is resistant to dilute hydrochloric and sulphuric acids, most organic acids, chlorine gas and chloride solutions. It is also resistant to alkalis. It combines with oxygen at red heat and chlorine at 550K.

Physical Information

Atomic Number	22
Relative Atomic Mass ($^{12}C=12.000$)	47.88
Melting Point/K	1933
Boiling Point/K	3560
Density/kg m^{-3}	4540 (293K)
Ground State Electron Configuration	[Ar]$3d^2 4s^2$
Electron Affinity(M-M$^-$)/kJ mol^{-1}	-2

Key Isotopes

nuclide	^{44}Ti	^{46}Ti	^{47}Ti	^{48}Ti
atomic mass	43.952	45.952	46.948	47.948
natural abundance	0%	8.2%	7.4%	73.8%
half-life	48 yrs	stable	stable	stable

nuclide	^{49}Ti	^{50}Ti
atomic mass	48.948	49.945
natural abundance	5.4%	5.2%
half-life	stable	stable

Ionisation Energies/kJ mol^{-1}

M - M$^+$	658
M$^+$ - M^{2+}	1310
M^{2+} - M^{3+}	2652
M^{3+} - M^{4+}	4175
M^{4+} - M^{5+}	9573
M^{5+} - M^{6+}	11516
M^{6+} - M^{7+}	13590
M^{7+} - M^{8+}	16260
M^{8+} - M^{9+}	18640
M^{9+} - M^{10+}	20830

Other Information

Enthalpy of Fusion/kJ mol^{-1} 20.9
Enthalpy of Vaporisation/kJ mol^{-1} 425.5

Oxidation States

main   Ti^{IV}
others Ti^{-I}, Ti^{O}, Ti^{II}, Ti^{III}

Covalent Bonds /kJ mol^{-1}

not applicable

Tungsten W

General Information

Discovery

Tungsten was discovered by J.J. and F. Elhuijar in 1783 in Vergara, Spain. However, in 1779 Woulfe examined the mineral wolframite and concluded it must contain a new element. An alternative name for tungsten is wolfram, from this discovery.

Appearance

Tungsten metal is silvery-white and lustrous, but the element is usually obtained as a grey powder.

Source

The principal tungsten containing ores are scheelite and wolframite. Commercially, the metal is obtained by reducing tungsten oxide with hydrogen or carbon.

Uses

Tungsten and its alloys are used extensively for filaments for electric lamps, electron tubes and television tubes. As it has the highest melting point of all metals it is used in numerous high-temperature applications. High-speed tool steels contain tungsten, as does a new "painless" dental drill which spins at ultra-high speeds.

Tungsten carbide is of great importance to the metal-working, mining and petroleum industries.

Calcium and magnesium tungstates are widely used in fluorescent lighting.

Biological Role

Tungsten has no known biological role, and has low toxicity.

General Information

Tungsten has the highest melting point and lowest vapour pressure of all metals, and at temperatures over 1650K has the highest tensile strength. The metal resists attack by oxygen, acids and alkalis.

Physical Information

Atomic Number	74
Relative Atomic Mass ($^{12}C=12.000$)	183.85
Melting Point/K	3680
Boiling Point/K	5930
Density/kg m^{-3}	19300 (293K)
Ground State Electron Configuration	[Xe]$4f^{14}5d^{4}6s^{2}$
Electron Affinity(M-M$^-$)/kJ mol^{-1}	119

Key Isotopes

nuclide	^{180}W	^{182}W	^{183}W	^{184}W
atomic mass	179.9	181.9	182.9	183.9
natural abundance	0.10%	26.3%	14.3%	30.7%
half-life	stable	stable	stable	stable

nuclide	^{185}W	^{186}W	^{187}W
atomic mass		185.9	
natural abundance	0%	28.6%	0%
half-life	75 days	stable	23.9 h

Ionisation Energies/kJ mol^{-1}

M - M$^+$	770
M$^+$ - M^{2+}	1700
M^{2+} - M^{3+}	2300
M^{3+} - M^{4+}	3400
M^{4+} - M^{5+}	4600
M^{5+} - M^{6+}	5900
M^{6+} - M^{7+}	
M^{7+} - M^{8+}	
M^{8+} - M^{9+}	
M^{9+} - M^{10+}	

Other Information

Enthalpy of Fusion/kJ mol^{-1} 35.2
Enthalpy of Vaporisation/kJ mol^{-1} 824.2

Oxidation States   W^{-IV}, W^{-II}, W^{-I}, W^{O}, W^{II}, W^{III}, W^{IV}, W^{V}, W^{VI}

Covalent Bonds /kJ mol^{-1}
not applicable

Uranium   *U*

General Information

Discovery

Uranium was discovered by M.H. Klaproth in 1789 in Berlin, Germany, and isolated by E.M. Peligot in Paris, France.

Appearance

Uranium is a radioactive, silvery metal.

Source

Uranium occurs naturally in several minerals such as pitchblende, uraninite and carnotite. It is also found in phosphate rock and monazite sands. It can be prepared by reducing uranium halides with Group I or Group II metals, or by reducing uranium oxides with calcium or carbon at high temperatures.

Uses

Uranium is of great importance as it provides nuclear fuel. Uranium is allowed to decay into plutonium under controlled conditions in a breeder reactor, where a chain reaction is set up and maintained. During this process, large amounts of energy are released. This energy is used to generate electrical power, to make isotopes for peaceful purposes, and to make nuclear weapons.

Biological Role

Uranium has no known biological role. It is toxic due to its radioactivity.

General Information

Uranium is malleable, ductile and tarnishes in air. It is dissolved by acids but not by alkalis. In a finely divided state it is pyrophoric.

Physical Information

Atomic number	92
Relative Atomic Mass ($^{12}C=12.000$)	238.03
Melting Point/K	1405.5
Boiling Point/K	4018
Density/kg m^{-3}	18950 (293K)
Ground State Electron Configuration	[Rn]$5f^3 6d^1 7s^2$

Key Isotopes

nuclide	^{234}U	^{235}U	^{236}U	^{238}U
atomic mass	234.04	235.04	236.05	238.05
natural abundance	0.005%	0.720%	0%	99.28%
half-life	2.47×10^5 yrs	7×10^8 yrs	2.39×10^7 yrs	4.51×10^9 yrs

Ionisation Energies/kJ mol^{-1}

M – M$^+$	584
M$^+$ – M^{2+}	1420
M^{2+} – M^{3+}	
M^{3+} – M^{4+}	
M^{4+} – M^{5+}	
M^{5+} – M^{6+}	
M^{6+} – M^{7+}	
M^{7+} – M^{8+}	
M^{8+} – M^{9+}	
M^{9+} – M^{10+}	

Other Information

Enthalpy of Fusion/kJ mol^{-1} 15.5
Enthalpy of Vaporisation/kJ mol^{-1} 417.1

Oxidation States

main   U^{VI}
others   U^{II}, U^{III}, U^{IV}, U^{V}

Covalent Bonds /kJ mol^{-1}

not applicable

Vanadium V

General Information

Discovery

Vanadium was discovered by A.M. del Rio in 1801 in Mexico City. However, a French chemist incorrectly declared that this new element was impure chromium, and del Rio accepted this judgement. Vanadium was rediscovered by N.G. Selfstrom in 1831 in Falun, Sweden.

Appearance

Vanadium is a shiny, silvery, soft metal.

Source

Vanadium is found in about 65 different minerals including vanadinite, carnotite and patronite, and also in phosphate rock, certain iron ores and some crude oils in the form of organic complexes.

Vanadium of high purity can be obtained by the reduction of vanadium trichloride with magnesium. Much of the vanadium metal now being produced is made by calcium reduction of vanadium (V) oxide in a pressure vessel.

Uses

About 80% of the vanadium produced is used as a steel additive. In this form it produces one of the toughest alloys for armour plate, axles, piston rods and crankshafts. Less than 1% of vanadium and as little chromium make steel shock- and vibration-resistant.

Vanadium (V) oxide is used in ceramics, as a catalyst and in producing superconducting magnets.

Biological Role

Vanadium is an essential trace element but some compounds are toxic.

General Information

Vanadium has good corrosion resistance to alkalis, dilute acids and salt water, but the metal oxidises rapidly above 660K.

The element was named after the Scandinavian goddess Vanadis because of its beautiful multi-coloured compounds.

Physical Information

Atomic Number	23
Relative Atomic Mass ($^{12}C=12.000$)	50.942
Melting Point/K	2160
Boiling Point/K	3650
Density/kg m^{-3}	6110 (292K)
Ground State Electron Configuration	[Ar]$3d^3 4s^2$
Electron Affinity(M-M$^-$)/kJ mol^{-1}	61

Key Isotopes

nuclide	^{48}V	^{49}V	^{50}V	^{51}V
atomic mass	47.952	48.948	49.947	50.944
natural abundance	0%	0%	0.250%	99.75%
half-life	16 days	330 days	6×10^{15} yrs	stable

Ionisation Energies/kJ mol^{-1}

M - M$^+$	650
M$^+$ - M^{2+}	1414
M^{2+} - M^{3+}	2828
M^{3+} - M^{4+}	4507
M^{4+} - M^{5+}	6294
M^{5+} - M^{6+}	12362
M^{6+} - M^{7+}	14489
M^{7+} - M^{8+}	16760
M^{8+} - M^{9+}	19860
M^{9+} - M^{10+}	22240

Other Information

Enthalpy of Fusion/kJ mol^{-1}	17.6
Enthalpy of Vaporisation/kJ mol^{-1}	459.7

Oxidation States

main   V^{III}, V^{IV}, V^{V}

others   V^{-III}, V^{-I}, V^{O}, V^{I}, V^{II}

Covalent Bonds /kJ mol^{-1}

not applicable

Xenon Xe

General Information

Discovery

Xenon was discovered by Sir William Ramsay and M.W. Travers in 1898 in London.

Appearance

Xenon is a colourless, odourless gas.

Source

Xenon is present in the atmosphere at a concentration of 0.086 parts per million by volume. It can be found in the gases which evolve from certain mineral springs. Commercially it is obtained by extraction from liquid air.

Uses

Xenon is little used outside research. However, it is used in certain specialised light sources which require an instant, intense light such as the high-speed electronic flash bulbs used by photographers. The high volatility of this element's electron structure produces this type of light. Xenon in a vacuum tube produces a beautiful blue glow when excited by an electrical discharge, and finds application in electron tubes, stroboscopic lights and bactericidal lamps.

Biological Role

Xenon has no known biological role. Xenon is not toxic, but its compounds are highly toxic because of their strong oxidising characteristics.

General Information

Xenon is inert towards most other chemicals but reacts with fluorine gas to form xenon fluorides. Xenon oxides, acids and salts are also known.

Physical Information

Atomic Number	54
Relative Atomic Mass ($^{12}C=12.000$)	131.29
Melting Point/K	161.3
Boiling Point/K	166.1
Density/kg m^{-3}	5.9 (gas, 273K)
Ground State Electron Configuration	$[Kr]4d^{10}5s^25p^6$
Electron Affinity(M-M$^-$)/kJ mol^{-1}	-41

Key Isotopes

nuclide	^{127}Xe	^{129}Xe	^{130}Xe	^{131}Xe
atomic mass		128.9	129.9	130.9
natural abundance	0%	26.4%	4.1%	21.2%
half-life	36.4 days	stable	stable	stable

nuclide	^{132}Xe	^{133}Xe	^{134}Xe	^{136}Xe
atomic mass	131.9		133.9	135.9
natural abundance	26.9%	0%	10.4%	8.9%
half-life	stable	5.27 days	stable	stable

Ionisation Energies / kJ mol^{-1}

M – M$^+$	1170.4
M$^+$ – M^{2+}	2046
M^{2+} – M^{3+}	3097
M^{3+} – M^{4+}	4300
M^{4+} – M^{5+}	5500
M^{5+} – M^{6+}	6600
M^{6+} – M^{7+}	9300
M^{7+} – M^{8+}	10600
M^{8+} – M^{9+}	19800
M^{9+} – M^{10+}	23000

Other Information

Enthalpy of Fusion/kJ mol^{-1} 3.1
Enthalpy of Vaporisation/kJ mol^{-1} 12.65

Oxidation States   Xe^{0}, Xe^{II}, Xe^{IV}, Xe^{VI}, Xe^{VIII}

Covalent Bonds / kJ mol^{-1}
Xe–O   84

Ytterbium Yb

General Information

Discovery
Ytterbium was discovered by J.C.G. de Marignac in 1878 in Geneva, Switzerland.

Appearance
Ytterbium has a bright, silvery lustre. It is soft, malleable and quite ductile.

Source
In common with many rare earth elements, ytterbium is found principally in the mineral monazite, from which it can be extracted by ion exchange and solvent extraction.

Uses
Ytterbium is little used outside research.

Biological Role
Ytterbium has no known biological role, and is non-toxic.

General Information
Ytterbium is slowly oxidised by the air, and reacts with water. It is readily attacked and dissolved by acids.

Physical Information

Atomic Number	70
Relative Atomic Mass ($^{12}C=12.000$)	173.04
Melting Point/K	1097
Boiling Point/K	1466
Density/kg m^{-3}	6965 (293K)
Ground State Electron Configuration	$[Xe]4f^{14}6s^2$
Electron Affinity(M-M$^-$)/kJ mol^{-1}	50

Key Isotopes

nuclide	^{168}Yb	^{169}Yb	^{170}Yb	^{171}Yb	^{172}Yb
atomic mass	167.9		169.9	170.9	171.9
natural abundance	0.14%	0%	3.06%	14.4%	21.9%
half-life	stable	31.8 days	stable	stable	stable

nuclide	^{173}Yb	^{174}Yb	^{175}Yb	^{176}Yb
atomic mass	172.9	173.9		175.9
natural abundance	16.1%	31.8%	0%	12.7%
half-life	stable	stable	101 h	stable

Ionisation Energies/kJ mol^{-1}

M − M$^+$	603.4
M$^+$ − M^{2+}	1176
M^{2+} − M^{3+}	2415
M^{3+} − M^{4+}	4220
M^{4+} − M^{5+}	
M^{5+} − M^{6+}	
M^{6+} − M^{7+}	
M^{7+} − M^{8+}	
M^{8+} − M^{9+}	
M^{9+} − M^{10+}	

Other Information

Enthalpy of Fusion/kJ mol^{-1} 9.2
Enthalpy of Vaporisation/kJ mol^{-1} 159

Oxidation States   Yb^{II}, Yb^{III}

Covalent Bonds /kJ mol^{-1}

not applicable

Yttrium   Y

General Information

Discovery

Yttrium was discovered by J. Gadolin in 1794 in Abo, Finland, and named after the Swedish village Yttria from which it was mined.

Appearance

Yttrium is a silvery-white, soft metal which is relatively stable in air due to formation of the oxide film.

Source

Yttrium occurs in nearly all the rare-earth minerals. It is recovered commercially from monazite sand and bastnaezite by reduction with calcium metal.

Uses

The largest use of yttrium is in the form of yttrium (III) oxide, which is used to produce phosphors which give the red colour in colour television tubes. It is also used in the making of microwave filters.

Yttrium is often used as an additive in alloys, and increases the strength of aluminium and magnesium alloys. It is also used as a detoxifier for non-ferrous metals. It has been used as a catalyst in ethylene polymerisation.

Yttrium-90, a radioactive isotope, has a dramatic medical use in needles which have replaced the surgeon's knife in killing pain-transmitting nerves in the spinal cord.

Biological Role

Yttrium has no known biological properties, and is non-toxic. It is a suspected carcinogen.

General Information

Yttrium reacts with water to give hydrogen. Finely divided metal is unstable in air, and metal turnings ignite in air, in contrast to lump metal which is stable in air.

Physical Information

Atomic Number	39
Relative Atomic Mass ($^{12}C=12.000$)	88.906
Melting Point/K	1795
Boiling Point/K	3611
Density/kg m^{-3}	4469 (293K)
Ground State Electron Configuration	[Kr]4d^{1}5s^2
Electron Affinity(M-M$^-$)/kJ mol^{-1}	-39

Key Isotopes

nuclide	^{88}Y	^{89}Y	^{90}Y
atomic mass	87.91	88.91	
natural abundance	0%	100%	0%
half-life	106.6 days	stable	64 h

Ionisation energies/kJ mol^{-1}

M - M$^+$	616
M$^+$ - M^{2+}	1181
M^{2+} - M^{3+}	1980
M^{3+} - M^{4+}	5963
M^{4+} - M^{5+}	7430
M^{5+} - M^{6+}	8970
M^{6+} - M^{7+}	11200
M^{7+} - M^{8+}	12400
M^{8+} - M^{9+}	14137
M^{9+} - M^{10+}	18400

Other Information

Enthalpy of Fusion/kJ mol^{-1}	17.2
Enthalpy of Vaporisation/kJ mol^{-1}	367.4

Oxidation States   Y^{III}

Covalent Bonds /kJ mol^{-1}

not applicable

Zinc   Zn

General Information

Discovery

Zinc was known in India and China before 1500.

Appearance

Zinc is a bluish-white, lustrous metal.

Source

Zinc is found in several ores, the principal ones being zinc blende, calamine and marmatite.

Commercially, zinc is obtained from its ores by concentrating and roasting the ore, then reducing it to zinc thermally with carbon or by electrolysis.

Uses

Zinc is used in alloys such as brass, nickel silver and aluminium solder. Large quantities of zinc are used to produce die castings which are important in the automobile, electrical and hardware industries. It is also used extensively to galvanise other metals such as iron to prevent rusting.

Zinc oxide is widely used in the manufacture of very many products such as paints, rubber, cosmetics, pharmaceuticals, plastics, inks, soaps, batteries, textiles and electrical equipment.

Zinc sulphide is used in making luminous dials and fluorescent lights.

Biological Role

Zinc is an essential trace element which is non-toxic but carcinogenic in excess. When freshly-formed zinc (II) oxide is inhaled a disorder called the "oxide shakes" or "zinc chills" can occur.

General Information

Zinc reacts with both acids and alkalis. It tarnishes in air. It is brittle at normal temperatures but malleable at 100-150K. It is a fair conductor of electricity, and burns in air at high red heat with the evolution of white clouds of the oxide.

Physical Information

Atomic Number	30
Relative Atomic Mass ($^{12}C=12.000$)	65.39
Melting Point/K	692.7
Boiling Point/K	1180
Density/kg m^{-3}	7133 (293K)
Ground State Electron Configuration	[Ar]$3d^{10}4s^2$
Electron Affinity(M-M$^-$)/kJ mol^{-1}	9

Key Isotopes

nuclide	^{64}Zn	^{65}Zn	^{66}Zn
atomic mass	63.929	64.926	65.926
natural abundance	48.6%	0%	27.9%
half-life	stable	243.6 days	stable

nuclide	^{67}Zn	^{68}Zn	^{70}Zn
atomic mass	66.927	67.925	69.925
natural abundance	4.1%	18.8%	0.6%
half-life	stable	stable	stable

Ionisation Energies/kJ mol^{-1}

M - M$^+$	906.4
M$^+$ - M^{2+}	1733.3
M^{2+} - M^{3+}	3832.6
M^{3+} - M^{4+}	5730
M^{4+} - M^{5+}	7970
M^{5+} - M^{6+}	10400
M^{6+} - M^{7+}	12900
M^{7+} - M^{8+}	16800
M^{8+} - M^{9+}	19600
M^{9+} - M^{10+}	23000

Other Information

Enthalpy of Fusion/kJ mol^{-1}	6.67
Enthalpy of Vaporisation/kJ mol^{-1}	114.2

Oxidation States

main	Zn^{II}
others	Zn^{I}

Covalent Bonds /kJ mol^{-1}

not applicable

Zirconium Zr

General Information

Discovery

Zirconium was discovered by M.H. Klaproth in 1789 in Berlin, Germany, and isolated by J.J. Berzelius in 1824 in Stockholm, Sweden.

Appearance

Zirconium is a hard, lustrous, greyish-white metal.

Source

Zirconium occurs in about 30 mineral species, the major ones being baddeleyite and zircon, found in Brazil. It is produced commercially by reduction of the chloride with magnesium.

Uses

Zirconium has very low absorption for neutrons, and is therefore useful in nuclear energy applications. More than 90% of zirconium production is used in this field, as reactors use many metres of zirconium alloy tubing.

Zirconium is exceptionally resistant to corrosion by most agents including sea water, acids and alkalis, and so is used extensively by the chemical industry where corrosive agents are in use.

With niobium, zirconium is superconductive at low temperatures and is used to make superconductive magnets.

Impure zirconium (IV) oxide is used for crucibles which will withstand heat shock, for furnace linings and by the glass and ceramics industries.

Biological Role

Zirconium has no known biological role. It is non-toxic.

General Information

The solid metal will burn in air, but with difficulty. When finely divided, however, it ignites spontaneously.

Physical Information

Atomic Number	40
Relative Atomic Mass ($^{12}C=12.000$)	91.224
Melting Point/K	2125
Boiling Point/K	4650
Density/kg m^{-3}	6506 (293K)
Ground State Electron Configuration	$[Kr]4d^2 5s^2$
Electron Affinity(M-M$^-$)/kJ mol^{-1}	43

Key Isotopes

nuclide	^{90}Zr	^{91}Zr	^{92}Zr	^{94}Zr
atomic mass	89.90	90.91	91.90	93.91
natural abundance	51.45%	11.32%	17.19%	17.28%
half-life	stable	stable	stable	stable

nuclide	^{95}Zr	^{96}Zr	^{97}Zr
atomic mass	94.91	95.91	
natural abundance	0%	2.76%	0%
half-life	65 days	3.6×10^{17} yrs	17 h

Ionisation Energies/kJ mol^{-1}

M - M$^+$	660
M$^+$ - M^{2+}	1267
M^{2+} - M^{3+}	2218
M^{3+} - M^{4+}	3313
M^{4+} - M^{5+}	7860
M^{5+} - M^{6+}	9500
M^{6+} - M^{7+}	11200
M^{7+} - M^{8+}	13800
M^{8+} - M^{9+}	15700
M^{9+} - M^{10+}	17500

Other Information

Enthalpy of Fusion/kJ mol^{-1}	23
Enthalpy of Vaporisation/kJ mol^{-1}	566.7

Oxidation States

main	Zr^{IV}
others	Zr^O, Zr^I, Zr^{II}, Zr^{III}

Covalent Bonds /kJ mol^{-1}

not applicable